Yongming He, Yingwu Chen
Imaging Satellites Task Planning

Also of Interest

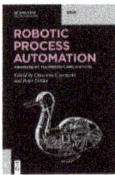

Yongming He, Yingwu Chen

Imaging Satellites Task Planning

Learning-Based BI-Level Models and Algorithms

DE GRUYTER

清華大學出版社
Tsinghua University Press

Authors

Assoc. Prof. Yongming He
College of Systems Engineering
National University of Defense Technology
P.R. China

Prof. Yingwu Chen
College of Systems Engineering
National University of Defense Technology
P.R. China

Contact via:
Tsinghua University Press Ltd.
Shuangqing Rd. Xue Yan Bldg. A-210
100084 Beijing
Haidian District
China
yuxx@tup.tsinghua.edu.cn

ISBN 978-3-11-158466-9
e-ISBN (PDF) 978-3-11-158510-9
e-ISBN (EPUB) 978-3-11-158519-2

Library of Congress Control Number: 2025931056

Bibliographic information published by the Deutsche Nationalbibliothek
The Deutsche Nationalbibliothek lists this publication in the Deutsche Nationalbibliografie;
detailed bibliographic data are available on the Internet at http://dnb.dnb.de.

© 2025 Walter de Gruyter GmbH, Berlin/Boston, Genthiner Straße 13, 10785 Berlin, and Tsinghua University
Press Ltd., China
Cover image: BlackJack3D / E+ / Getty Images
Typesetting: VTeX UAB, Lithuania

www.degruyter.com
Questions about General Product Safety Regulation:
productsafety@degruyterbrill.com

About the book

This book creatively puts forward the theory and application of learning bilevel task planning for imaging satellites. Starting from the development trend of task planning for imaging satellites, it elaborates on the design method of the task planning system for imaging satellites, the learning-based bilevel task planning model, and the solution framework, and introduces the operation control and task planning process of imaging satellites in-depth, to facilitate readers' quickly understanding of the research content of this book. For two problem-solving phases, deterministic algorithms and reinforcement learning algorithms are designed respectively, aiming to realize the organic unity between versatility and high efficiency during the solving process and between solution efficiency and accuracy. The research results of this book not only have practical applications in engineering problems but also contribute to the field of combinatorial optimization, providing new methods and ideas for solving other issues in this area.

This book is a comprehensive resource that bridges the gap between theory and practice in the field of satellite task planning. It can be a reference for senior undergraduates, graduate students, and college teachers in disciplines such as systems engineering, management science and engineering, aerospace engineering, operations research, artificial intelligence, etc. Engineers and technicians in the aerospace industry sector, researchers in related scientific research institutes, and scientists and technologists interested in satellite task planning will find this book a valuable tool to enhance their understanding and application of task planning principles.

https://doi.org/10.1515/9783111585109-201

Preface

Imaging satellite platform and load capacity has been gradually strengthened, while their number is increasing explosively, so that imaging satellites can achieve a broader range of applications and generate more significant social benefits, but also the imaging satellite task planning has brought new challenges: the refinement of the integrated control, resulting in an increase in the number of variables in the problem, decision-making dimensions, the solution space has become more prominent, and the requirement of result quality is higher; the normalization of the rapid response, resulting in the continuous enhancement of users' expectations for the time efficiency of imaging products, which puts forward higher requirements for the algorithm's computational efficiency and stability; the complexity of the constraints, resulting in the deepening of the coupling relationship between the variables in the problem, and the algorithm is required to achieve unified access and organic integration of different types of imaging satellites in order to enhance the comprehensive control capability and the effect of integrated application. Because of the above changes and needs, the contradiction between generalization and efficiency between solution efficiency and solution accuracy in the task planning process of imaging satellites has become increasingly acute. In view of this, the book introduces the theory and application of imaging satellite learning-based bilevel task planning. The main research content is developed into the following five aspects:

(1) A general approach to the design of imaging satellites task planning systems is presented

Focusing on the characteristics of operation and control processes of novel imaging satellites, the system design requirements are analyzed, the overall structure of the imaging satellite task planning system is sorted out, the business processes of imaging satellite operation and control are divided, and the interrelationships among the modules of the task planning system are abstracted. Focusing on the system design concept, we determine the overall ideas and principles. Taking this as the basic guideline, object-oriented visual modeling technology is used to realize the system's detailed design to ensure the design's consistency, the independence of the modules, and the agility of the development. The research content follows the software design specifications and fundamental principles, providing a referable and easy-to-implement methodological foundation for related researchers.

(2) A bilevel optimization model for the imaging satellite task planning problem is proposed

Based on the sorting and analysis of the operation and control process of imaging satellites, the standardized description and basic assumptions of the imaging satellite task planning problem are determined. Based on the overview of the current research sta-

https://doi.org/10.1515/9783111585109-202

tus of the imaging satellite task planning model and the analysis of the essential elements of the imaging satellite task planning problem, a problem decomposition scheme and a bilevel combinatorial optimization framework for imaging satellite task planning are proposed: the task assignment process selects a suitable visible time window for a task, while the task scheduling process decides the specific execution time of each task based on the determined time window. Under the learning-based bilevel combinatorial optimization framework, a mathematical planning model and a finite Markov decision model are designed for the task scheduling process and the task assignment process, respectively, and a learning-based bilevel task planning method integrating reinforcement learning and deterministic algorithms is proposed to give full play to the advantages of reinforcement learning and deterministic algorithms. This part of the research is the fundamental basis for analyzing and solving the task scheduling process and the task assignment problems. It provides a methodological foundation for realizing the unified modeling and solution of different satellites and scenarios.

(3) Two deterministic algorithms for solving the task scheduling problem are proposed

The task scheduling problem is considered as a mathematical planning model. The constraints in the model are divided into four categories. A unified constraint representation approach and a constraint-checking algorithm based on the timeline are designed to realize the efficient processing of complex constraints. According to the characteristics of the problem and the advantages of the deterministic algorithm, this part designs the heuristic algorithm based on the density of the residual tasks and the dynamic programming algorithm based on the task sequencing, and the design of the solution process ensures the stability of the results. The complexity analysis process illustrates the computational efficiency of the algorithms, while the optimality of the two algorithms can be proved through theoretical derivation when they satisfy different specific conditions, so as to ensure the quality of the algorithms' solutions. In addition, the computing process of these two algorithms is less coupled with the constraints, which improves the generality of the algorithms in the imaging satellite task scheduling problem. Simulation experiments verify that the heuristic algorithm based on residual task density has apparent advantages in time efficiency and arithmetic stability, and at the same time, the solution quality is no less than that of the three state-of-the-art comparative algorithms; the algorithm of the dynamic programming algorithm based on task sequencing obtains a higher task completion rate and task profit rate than the heuristic algorithm based on residual task density in an acceptable time, especially in the oversubscribed scenario. The task completion rate and task profit rate are improved by about 26 % and 19 % on average over the adaptive large neighborhood search algorithm, respectively.

(4) An improved deep Q-learning algorithm to solve the task assignment problem is proposed

Based on the finite Markov decision model, the design of each element of the model is further refined: taking into account the characteristics of the task assignment problem, such as numerous input parameters, complex relationships, and low information density, the action space and state space are narrowed as much as possible based on ensuring the completeness of the action space and the state space; meanwhile, the computation of the short-term return function and the representation of the value function are designed with the domain knowledge, which can effectively alleviate the problems of low training efficiency caused by sparse return during the training process. Based on this model, an improved deep Q-learning algorithm is designed, which includes a solution framework oriented to random initial states and an action-pruning strategy based on domain knowledge and constraints to improve the training efficiency of the algorithm. In the simulation experiments, the ablation of the algorithm is studied. The influence of the value function and attribute configuration in the algorithm and the integration of multiclass algorithms on the efficiency and effectiveness of the solution are discussed. The optimal configuration scheme of the activation function, loss function, and optimizer in the deep Q-learning algorithm for solving the task assignment problem of imaging satellites is determined. Meanwhile, the feasibility of the deep Q-learning algorithm in solving the task assignment problem and the superiority of integrating reinforcement learning and deterministic algorithms in the task planning problem is proved by analyzing the training and application process of the algorithm.

(5) Research results are validated in "SuperView-1" task planning scenario

Starting from the real-world project, the task planning system of "SuperView-1" is designed, and the rationality of the system is ensured by the design of the system's internal and external interfaces and data structure. A bilevel optimization model and an integrated planning algorithm for the "SuperView-1" constellation are established and used to solve the corresponding task-planning problems. The system's design, model, and algorithm follow the operation and control process of the constellation and the industry standard, which can be well connected to the practical application. Simulation experiments in 14 groups of daily task planning scenarios of the "SuperView-1" show that the solution accuracy of the two integrated algorithms designed in this book is better than that of the comparison algorithms in all scenarios and that they can solve the most experimental scenarios with limited computational resources, which demonstrates the high computational efficiency of the proposed integrated algorithms. Facing future complex application scenarios, the proposed integrated algorithms have significant advantages and potential.

This book is a collection of the leading academic achievements accumulated by the author during his study and work at the School of Systems Engineering, National University of Defense Technology (NUDT) from 2014 to 2023. Every word of this book could not be separated from the careful guidance of the authors' supervisor, Professor Yingwu Chen, as well as the support of fellow teachers and students in the group (special thanks to Xia Yang, Gao Yuanliang, Qiu Haodong, Mei Xuan, and Zhou Qinghua for their efforts in this book), colleagues in the aerospace industry, and the diligent efforts of editors in Tsinghua University Press for the publication of this book. In addition, this book has been supported by the National Natural Science Foundation of China Youth Science Foundation Project (72201273). The authors would like to express their heartfelt thanks to the experts, scholars, teachers, and students who have given their support and help with this book!

This book carries out preliminary research work around the imaging satellite learning-based bilevel task planning system, model, method, application, etc., which has high academic and application value, and many of these scientific issues deserve further in-depth research, and the authors are eager to stimulate the interest of the general readers in the related scientific issues throughout this book. Although the authors have done their best to improve the quality of the content, due to the limited level, the book inevitably still has errors and worth improving. We sincerely welcome your valuable comments. Thank you very much!

Changsha, China   The authors
March 2025

Contents

1 Introduction

The rapid development of the Chinese satellite industry has led to an increasingly urgent need for efficient spacecraft management and intelligent task planning. Especially in the past two decades, the double leap in the number and quality of Chinese imaging satellites has led to a steep increase in the complexity of the task planning problem [1]. How to rationally allocate satellite resources to maximize their practical benefits has gradually become one of the hot issues in the field of combinatorial optimization and aerospace. This chapter starts with the basic concepts of imaging satellites and imaging satellite task planning, analyzes the characteristics and difficulties of the imaging satellite task planning problem, and analyzes the realistic needs of the imaging satellite task planning problem, so as to draw out the motivation and significance of the research content of this book, and then summarizes the book's characteristics and innovations.

1.1 Imaging satellites

As an important class of space-based information acquisition tools, imaging satellites acquire remote sensing data from a space perspective through their imaging payloads. Technical departments further process these data to produce valuable imagery products and distribute them to users [2]. Due to the unique nature of the satellite's working environment, imaging satellites have advantages in many practical application scenarios that are incomparable to ground-based information acquisition tools, providing efficient and reliable information support for work and production activities in many fields such as national defense, economy, and society, which greatly enhancing social productivity and national defense strength [3]. In recent years, with the continuous breakthroughs in imaging payload and platform technology, the number of satellites has grown exponentially, and the application areas of imaging satellites have also deepened from the "roughly plan" to the "precisely decision" [4, 5]: the increasing popularity of satellites with submeter resolution has led to the application of imaging satellites in scenarios [6]; the ability of satellites to maneuver in an attitude has been strengthened gradually, and the flexibility of satellites in performing their operations, making various types of actions are also becoming diverse, enabling the satellites to accomplish more imaging tasks within a limited time [7, 8, 9]; the increasing scale of satellites operating in orbit enhances the ability of the imaging satellite system to work around the clock, making the application modes and scenarios of the imaging satellites further extending [10].

As an important part of the large family of remote sensing satellites, the main work of imaging satellites is based on imaging payloads. Typical imaging payloads include optical payloads, synthetic aperture radar (SAR) payloads, infrared imaging payloads, etc., which receive and process signals through different types of sensors. These signals are usually converted into electrical signals and then analyzed into remote sensing image

https://doi.org/10.1515/9783111585109-001

information [8] in conjunction with signal processing and related techniques. The available imaging area that a satellite can acquire is determined by the width of the payload's field of view, the satellite's orbit, and the satellite's attitude maneuvering capability [11]. The schematic of satellite imaging is shown in Figure 1.1.

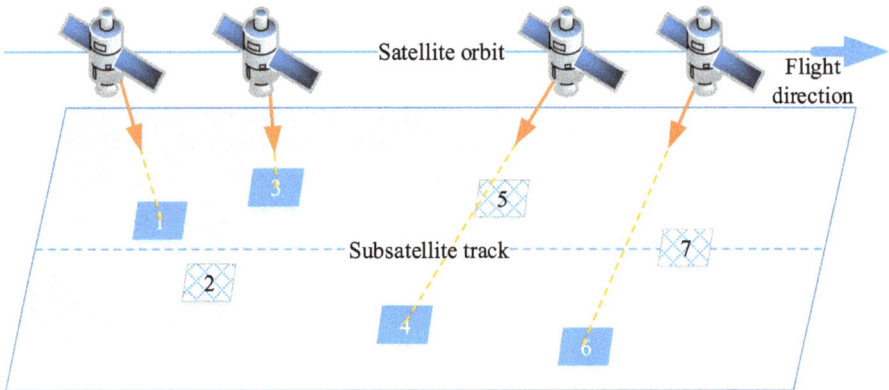

Figure 1.1: Schematic diagram of satellite imaging.

1.1.1 Category

The basic working principles of imaging satellites are similar, but according to different criteria, imaging satellites can be classified according to their purpose, type of orbit, type of payload, attitude maneuvering capability, and mode of control.

(1) According to purpose, imaging satellites can be categorized into reconnaissance satellites, early warning satellites, mapping satellites, meteorological satellites, and so on. Among them, the function of reconnaissance satellites is to reconnaissance by their high-resolution imaging payloads, identify and track targets on the ground, sea surface, or in the sky, and to obtain as detailed target information as possible; the primary function of early-warning satellites is to detect and monitor potential strategic threats, and to provide reliable intelligence support for their own strategic defense and counterattacks. The surveying and mapping satellites mainly use wide-format imaging payloads to realize rapid and seamless image coverage in the region and provide services in the fields of land resources planning, agricultural and forestry planning, urban transportation planning, etc. The meteorological satellites collect and process all kinds of meteorological information by carrying various kinds of meteorological remote sensors, which are widely used in environment monitoring, disaster prevention and mitigation, and atmospheric science and other researches.

(2) According to orbital altitude, imaging satellites can be categorized into low Earth orbit (LEO) satellites, medium Earth orbit (MEO) satellites, high Earth orbit (HEO) satellites, geostationary orbit (GEO) satellites, elliptical orbit satellites, and so on. Low-orbit

satellites usually refer to satellites with an orbital altitude of less than 2,000 km, and at present the vast majority of imaging satellites are designed to work on the low orbital plane, with the advantage of short data transmission delay and the same imaging payloads that can obtain higher-resolution data products; high-orbit satellites usually refer to satellites with an orbital altitude of more than 20,000 km, and one class of special high-orbit satellites is the geostationary orbit satellites, with an orbital altitude of 36,000 km. High-orbit satellites have a large imaging coverage, but their resolution is usually lower than that of low-orbit satellites, and they also need more powerful supporting equipment to cooperate with them to complete imaging and data reception. Medium orbit satellites and elliptical orbit satellites combine the advantages and disadvantages of LEO and HEO satellites, and the corresponding satellite orbits are flexibly designed according to actual needs.

(3) According to payload type, imaging satellites can be classified into optical imaging satellites, infrared imaging satellites, synthetic aperture radar (SAR) imaging satellites, and so on. Optical imaging satellites are a class of satellites that mainly use visible light payloads to realize imaging, and their advantages in data intuition, low difficulties of post-processing, and high resolution make them the most widely used class of imaging satellites, but the imaging effect of optical imaging satellites is easily affected by light conditions, and they cannot obtain effective image products after being blocked by features or clouds; infrared imaging satellites and SAR satellites synthesize images by means of infrared radiation and microwave radar signals, respectively, and their advantage is that they are less affected by light and climate.

(4) According to attitude maneuvering capability, imaging satellites can be classified into agile satellites and nonagile satellites. The agility of satellites is one of the characteristics of novel imaging satellites. Traditional nonagile satellites can only image the target at a specific point in time due to the fact that the pitch angle of the satellite platform and imaging satellite payload cannot be changed during the satellite flight. With the development of satellite platform and payload technology, the satellite has a strong attitude maneuvering ability, which means the satellite can be in a certain time interval to decide the order of implementation of the imaging task and the actual implementation of the moment to enhance the efficiency of the imaging satellite and the flexibility of the task planning scheme. If the imaging satellite has the above ability, it is said that the satellite has agility. The agility of imaging satellites, on the one hand, improves the efficiency of satellite utilization and, on the other hand, brings a more complex class of constraints (time-dependent constraints [12]) to the control process of imaging satellites. Satellite agility is shown in Figure 1.2.

(5) According to control mode, imaging satellites can be categorized into autonomous and nonautonomous satellites. With the improvement of the satellite platform and payload capacity, imaging satellites gradually have the conditions to change from the "command executor" to the "decision maker" with a certain degree of intelligence. Autonomous satellites can dynamically adjust their task planning schemes according to on-board power, storage, and received task information, thus realizing

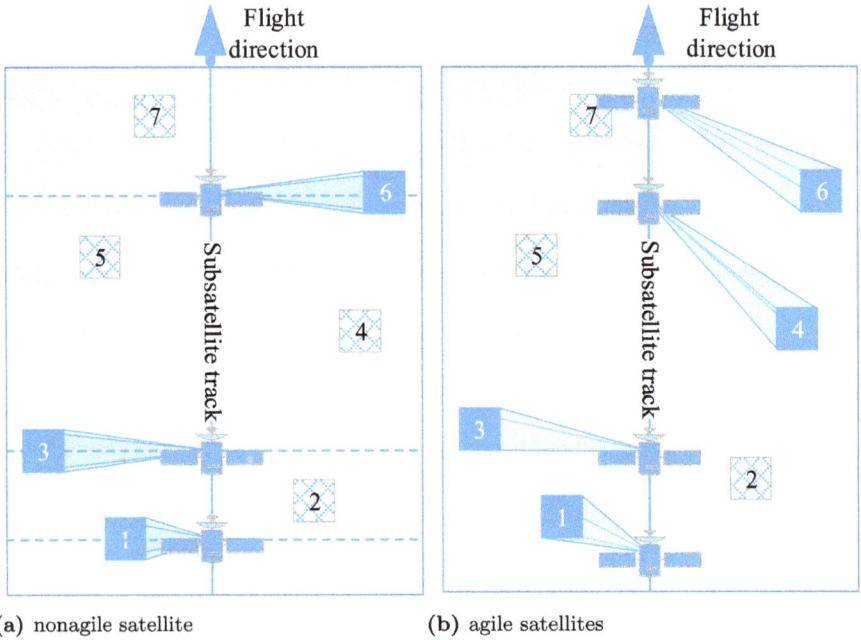

(a) nonagile satellite   (b) agile satellites

Figure 1.2: Schematic of the agility of imaging satellites.

rapid response to unexpected events and improving the actual utilization rate of satellite resources.

1.1.2 Trends and challenges

With the rapid development of imaging satellite-related technologies in China, high-performance imaging satellites have sprung up to meet the increasingly urgent and complex needs for intelligence assurance and land resources monitoring. The new imaging satellites are equipped with brand-new technical features and application modes, which expand the connotation and extension of intelligent decision-making related technologies in the satellite industry and also bring new challenges to the design of intelligent task planning systems and modeling algorithms.

(1) The development of new imaging satellites is the material basis for upgrading the ability to perform complex missions
With the rapid development of satellite platforms and payload technologies, satellite platforms have become more complex in terms of control and task modes. The launch of landmark satellites such as the "Gaofen Duomo (GFDM)" and the "Shiyan 20C (SY-20C)" represents China's progress from a major spacefaring country to a space power. The new imaging satellites have made many previously impossible functions possible: complex

tasks such as stereoscopic imaging, imaging in motion, and nontracking push scanning can be accomplished directly by the new satellites. At the same time, the completion of each complex task is accompanied by more influencing factors and more complex constraints, such as a particular type of agile satellite, e. g., a regional target of the real-time imaging task, the need for dozens of actions, and these actions exist with many complex logical relationships between them, so that in the implementation of these tasks in the process of not breaking through the objective conditions of the limitations. These actions are arranged according to certain rules and constraints, and an error in any one of them will affect the execution of all subsequent actions. In addition, the enhancement of satellite hardware capabilities has also brought great risks to satellite health management in orbit, requiring more precise methods of status monitoring, which in turn has led to increased pressure on data processing in the task planning process.

(2) The application of new imaging satellites is an effective way to alleviate the pressure of data transmission and processing

By the end of 2022, the number of satellites in orbit in China is close to 600, of which the number of satellites in orbit in the field of remote sensing satellites also exceeds 200, ranking among the top in the world. New application modes, such as satellite data relay transmission, have eased the bottleneck problems of traditional satellites limited by measurement and control resources, but inefficient task planning also results in a huge waste of resources, leading to a significant reduction in the comprehensive effectiveness of satellites. On the other hand, the current task planning process still relies on the experience and skills of the relevant staff, which asks operators of satellite management departments to understand the operational details of different satellites and payload use methods. Due to its low fault tolerance, the task planning process is very cumbersome. It usually requires operators to put in a lot of time and effort to learn and understand the basic knowledge of satellites and the application process, and the cost of learning is very high; in the process of operation and security, operators also often need to communicate with other satellites. In the process of operation and security, the operators often need to proofread and confirm the instructions and parameters with the satellite developer, with high labor cost and high risk of error. This is obviously not suitable for the future development of large-scale satellite fleets, and there is an urgent need to adopt intelligent means to enhance the efficiency of satellite management and control.

(3) Intelligent satellite task planning is key to developing space-based rapid response capabilities

The use of imaging satellites to cope with major natural disasters, public emergencies, the fight against crime, and regional hotspot events has great potential. However, these events have a high degree of spontaneity and uncertainty in time and space. The use of imaging satellites to obtain relevant intelligence support puts forward high requirements for the satellite's work efficiency and precision. At present, the imaging satellite task planning cycle usually ranges from one day to several days, and after the planning scheme has been formulated, the corresponding satellite commands are generated and injected into the on-orbit satellites within the time window of the satellite and the ground. The satellites will strictly follow the scheme to implement the observation of the ground. In addition to the objective factors of hardware and management, another important reason for the long planning cycle is that the task planning algorithm is not accurate and fast enough to flexibly select appropriate planning strategies in different scenarios, which leads to the planning scheme not being able to satisfy the demand. It needs to go through approval by satellite managers and then improve the quality of the scheme, which elongates the progress of the whole process of satellite operation and control. Therefore, on the one hand, it is necessary to reorganize the operation and control mode of imaging satellites, including improving the content of data transmission in the measurement and control process, realizing the change from basic "command level" information transmission to "mission and task level" and striving to realize the process of interaction and synchronization of the information between the on-board system and the ground systems with the least amount of data. On the other hand, it is necessary to improve the versatility of the algorithm so that the operation and control process of imaging satellites output stable and satisfactory task planning schemes in different scenarios and reduce repetitive work.

1.2 Imaging satellites task planning

Imaging satellite task planning (referred to as "task planning" or "planning" in this book) is the core decision-making process of the operation and control process of imaging satellites, and is the "brain" of the satellite planning system. It needs to take into account the capabilities and constraints of all resources related to satellite operation and control (e. g., satellite resources, operations resources, measurement and control resources, data transmission resources, etc.), and utilize scientific computation methods to generate a reasonable imaging satellite task planning scheme [13]. The satellite task planning scheme mainly includes the work scheme of the controlled imaging satellite and the data transmission schemes of the relevant resources, and also considers the feasibility of the measurement and control scheme. Therefore, the imaging satellite task planning process must follow two main principles:

(1) Must ensure safe and stable operation of each resource
Imaging satellites are expensive, and all constraints must be strictly met. Uncertainty in the resulting planning scheme must be avoided as much as possible. Errors in the control of imaging satellites that lead to abnormal satellite operation are usually accompanied by huge costs, so the task planning process in actual engineering projects avoids the use of randomized algorithms to ensure that the results of the calculations are interpretable and stable.

(2) Maximize the overall operational efficiency of the satellite system
The algorithm is required to take into account the actual situation of each resource in the operations process, improving the level of speed, accuracy, and stability in the overall calculation process of task planning, and fully exploit the imaging satellite's information acquisition capability, to ensure the computational efficiency and solving effect of the imaging satellite task planning process in all kinds of complex environments.

A wide range of scholars have considered the imaging satellite task planning problem as a combinatorial optimization problem and have solved it by specific mathematical planning algorithms [9], taking into account the specific constraints of each satellite. It also transforms the problem from a "soft management" problem relying on the domain experience of managers or decision makers to a scientific problem in the field of operations research and optimization: applying optimization theory and intelligent computational methods to find a satisfactory task planning scheme using limited computational resources in the imaging satellite task planning problem with different conditions, and providing managers and decision makers with a basic guiding for decision-making, to improve the operation efficiency of the imaging satellite operation control system. The basic form of combinatorial optimization model for imaging satellite task planning problem is shown in Model (1.1):

$$
\begin{aligned}
\text{Minimize} \quad & F(\boldsymbol{r}, \boldsymbol{es}) \\
\text{Subject to} \quad & G_k(\boldsymbol{r}, \boldsymbol{es}) \leq 0, \quad k = 1, \dots, g \\
& [\boldsymbol{r}, \boldsymbol{es}] \in \Omega
\end{aligned}
\tag{1.1}
$$

where the objective function $F(\boldsymbol{r}, \boldsymbol{es})$ is used for evaluating the goodness of the scheme, where $\boldsymbol{r}$ denotes the vector consisting of the imaging resources corresponding to the completion of the tasks, and $\boldsymbol{es}$ denotes the vector consisting of the start time of the execution of all the tasks; $G_k(\boldsymbol{r}, \boldsymbol{es}) \leq 0$ represents the constraint k of the imaging satellite task planning problem; g denotes the number of constraint entries in the problem; $[\boldsymbol{r}, \boldsymbol{es}]$ denotes the matrix formed by the decision variables $\boldsymbol{r}$ and $\boldsymbol{es}$, which represents a solution; Ω represents the feasible domain of the solution. In this model, the decision variables $\boldsymbol{r}$ and $\boldsymbol{es}$ in the imaging satellite task planning problem are vectors consisting of natural numbers, and there is a deep logical relationship. Considering the context and characteristics of the specific problem, the vast majority of combinatorial optimization models for imaging satellite task planning problems have been proven to be NP-hard. The optimal solution cannot be guaranteed in polynomial time for such problems, a typical of which is the Agile Earth Observation Satellite Scheduling Problem (AEOSSP) [14].

It is easy to see that this class of models can be considered as two solution processes [15] depending on the decision variables:
① On which resource is each task executed? (task assignment process).
② When exactly is each task executed? (task scheduling process).

If one focuses only on the decision-making process itself and does not discuss specific constraints, real-life combinatorial optimization problems that can be described to take shape as Model (1.1) are prevalent in many industries [16] (e. g., production, manufacturing, transportation, logistics, aerospace, military operations, etc.), such as the job-shop scheduling problem (JSP) [17], the traveling salesman problem with time window (TSPTW) [18, 19], vehicle routing problem (VRP) [20, 21] and so on. These problems have strong theoretical research value and application prospects, so they have attracted the attention of many experts in management science, system science, operations research, and other related fields, and the solution methods designed for these classes of problems have sprung up in recent years.

The imaging satellite task planning problem was often mapped to a classical problem described above and solved by applying the corresponding state-of-the-art algorithms [22, 23]. This approach is conducive to the standardization of the problem model and the proceduralization of the solution process. However, the steady improvement of the hardware level of imaging satellites, the continuous innovation of the control methods, and the in-depth thinking of the satellite control department on the satellite application mode, all these changes have made the imaging satellite task planning problem increasingly complicated. It is steeply tricky to map the imaging satellite task planning problem into a classical problem, in other words, the classical problem may not be able to well describe the essential characteristics of the imaging satellite task planning problem under complex multiconstraint conditions. Functional positioning, platform and payload capabilities, space location, and control methods of each imaging satellite are significantly different, which leads to the nature of its corresponding task planning problem varies [24, 25, 26]. Figure 1.3 summarizes the main factors that need to be considered to build a general task planning model for imaging satellites.

The nature of the problem may change by modifying only one constraint, or even by changing the range of values of the constraint. For example, when considering satellites with agility, the imaging satellite task planning problem is an NP-hard problem, otherwise it is not [27]. It is precisely the diversity of objective conditions in the problem, such as user requirements, operating environment, payload characteristics, and control modes, that leads to the variety of the imaging satellite task planning model, which in turn leads to a further widening of the gap between the problems discussed in practical engineering and theoretical studies: many algorithms that perform well in classical combinatorial optimization problems are poorly or even inapplicable to solve the imaging satellite task planning problem in practical engineering. Specifically, compared with the classical combinatorial optimization problem, the imaging satellite task planning problem has the following three distinctive features and difficulties:

(1) The analytic nature of the imaging satellite task planning problem is difficult to be described in a uniform way, which makes it difficult to manually design reasonable solution rules for the problem [28, 29]. Since imaging satellite task planning is an important part of a complex satellite operation and control system, the imaging requirements proposed by the user need to undergo a series of complex operations before they can be transformed into the input parameters of the imaging satellite task planning model [30]. These operations are too complex to be expressed by simple functions or

Single-satellite task planning				
Purpose	Reconnaissance	Warning	Mapping	Meteorological	······
Payload	Optical	Infrared	SAR	······
Agility	Agile	Semi-agile	Non-agile	······
Orbit	LEO	MEO	HEO	GEO	······
Control mode	Autonomous	Semiautonomous	Nonautonomous

Capacities and Constraints

Multi-satellite task planning				
Constellation configuration	Single type, single orbit	Multi type, single orbit	Single type, multi orbit	Multi type, multi orbit

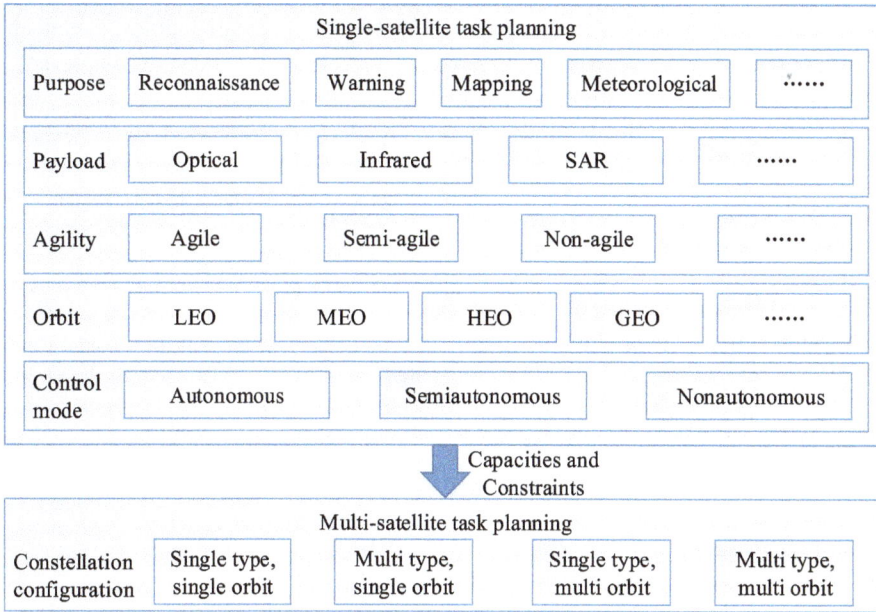

Figure 1.3: Factors of modeling imaging satellite task planning problems.

formulas, which brings great resistance to mining the intrinsic features of the problem model and summarizing the solution rules.

(2) The constraints of the imaging satellite task planning problem are more complex than classical problems, and thus, it becomes more and more challenging to design algorithmic rules for specific constraints [31]. The constraints of the imaging satellite task planning problem are organized with reference to the satellite usage documents provided by the satellite manufacturer and the satellite management and control party, so the constraints of a satellite in actual engineering can be as many as dozens or hundreds [31]. On the other hand, the constraints in the classical planning model are generally numerical constraints and rarely contain logical constraints; the constraints of the imaging satellite task planning problem do not necessarily change linearly with the parameters (e. g., absolute value constraints, quadratic term constraints, etc.), and they may even be more complex logical constraints (e. g., uniqueness constraints, nonempty constraints, etc.). The complex form of constraints brings great difficulties to the model simplification and analysis, which requires substantial skills in operations research, as well as an in-depth understanding of the nature of the problem in order to be accomplished.

(3) The objective function of the imaging satellite task planning model is more diversified than that of classic problems [32, 33]. This is mainly based on the application scenarios or missions oriented to each imaging satellite system, combined with the preferences of decision makers to determine jointly. Different application goals in the real-

world problem lead to different optimization objectives in the mathematical model, under which the rules and strategies in the task planning algorithm need to be tailored in order to be able to use as little computing time as possible to obtain a satisfactory task planning scheme. The influence of the decision-maker's preferences on the model and the solution process is less considered in classical combinatorial optimization problems.

As the imaging satellite task planning problem and model have the above characteristics and difficulties, the current industrial sector to solve the imaging satellite task planning problem basically still stays in the situation of "one satellite, one system," i.e., according to the design scheme of the imaging satellite and the constraints on the use of the satellite to build the satellite task planning system as well as corresponding models and algorithms. It is difficult to achieve standardized design and unified management of imaging satellite task planning systems, models, and algorithms. Meanwhile, imaging satellite task planning algorithms will also face more significant challenges: as the complexity of the problem increases, the contradiction between solving efficiency and solving accuracy of algorithms will become more prominent.

With the rapid growth in the number of satellites and the increasing complexity of the satellite operating environment and constraints, the continued use of this model for the development of an imaging satellite task planning system not only requires the investment of more costs for the development and maintenance of the system and algorithms, but also makes it more difficult to ensure the reliability and stability of the developed system. The main reason why it is difficult to achieve unified development of an imaging satellite task planning system is that imaging satellites are constantly updated iteratively, which makes it difficult to describe the constraints and objective functions of imaging satellite task planning problems in a unified way. Therefore, in order to reduce the pressure on the development and maintenance of imaging satellite operation and control system and improve the overall application efficiency of imaging satellite systems, it is imperative to study imaging satellite task planning models that can uniformly describe and deal with various types of constraints and design stable and efficient algorithms.

1.3 Motivation and significance

This study focuses on alleviating two increasingly prominent contradictions in imaging satellite task planning problems by taking into full consideration the commonalities and characteristics between imaging satellite task planning problems and classical combinatorial optimization problems: the contradiction between the generality and validity of the task planning models, and the contradiction between the computational efficiency and accuracy of solution algorithms. We are committed to exploring a more general and efficient system for solving the imaging satellite task planning problem. Based on the understanding of the system design scheme, the imaging satellite task planning model is proposed to unify the description of various elements, maximize the decoupling of different backgrounds and constraints in specific problems with the decision-making process, and get rid of the current development status of "one satellite, one system," so as to shorten the development cycle and maintenance cost of the task planning system in the satellite industry. On the basis of this model, reasonable algorithms are designed to enhance the universality, intelligence, and rapid response of the satellite task planning algorithms under the premise of guaranteeing the solution accuracy, so as to ultimately realize the planning process of imaging satellites in various kinds of complex and changeable environments in a quick and good way.

The imaging satellite task planning problem has long been described as an optimization model, and either deterministic algorithms or stochastic algorithms have been designed to solve it [30, 34]. Deterministic algorithms are algorithms in which there are no random variables in the algorithmic process, and the decision-making is based on a deterministic strategy. Dynamic programming algorithms [35] and branch and pricing algorithms are typical deterministic algorithms. In contrast, stochastic algorithms such as genetic algorithms, ant colony algorithms, simulated annealing algorithms, and other intelligent optimization algorithms apply random parameters to achieve the algorithm's search process in the solution space. The advantages and disadvantages of these two types of algorithms are also undeniable: the convergence direction of the deterministic algorithm is straightforward and does not require repeated iterations, so its computational efficiency is usually higher; stochastic algorithms seem to realize the generality of the solution process, but due to the nature of stochastic algorithms, it is difficult to reconcile the contradiction between its solution quality and its solution efficiency. In addition, the solution quality of stochastic algorithms is not as stable as that of deterministic algorithms, which is contrary to the first principle introduced in Section 1.2. Therefore, deterministic algorithms are preferred to the more popular intelligent optimization algorithms for solving imaging satellite task planning problems in real-world applications.

However, in the face of complex problem contexts and large-scale application scenarios, deterministic algorithms usually also have difficulty in guaranteeing the optimality of the algorithms in an acceptable time: in the context of the increasing complexity of the problem, if the imaging satellite task planning problem is simply built in the form of Model (1.1) and a single deterministic algorithm is used to solve the problem, which results in the complexity of the mathematical model and a decrease in the versatility of the model. Meanwhile, the increasing difficulties of simplifying and deforming the model lead to an increase in the cost of designing algorithms for solving the model and difficulty in guaranteeing the efficiency and the quality of the algorithms designed [27, 36]. Therefore, based on the current and future development level of the satellite industry, this solution idea is not conducive to the theoretical research and large-scale popularization and application of the imaging satellite task planning problem.

As a class of paradigms in the field of machine learning that specializes in decision-making problems, Reinforcement Learning (RL) has gradually attracted the attention of experts in the field of planning and scheduling in recent years, and has been successfully applied to classical scheduling problems such as JSP, VRP, and so on [37, 38, 39]. The basic principle of reinforcement learning is that by studying the relationship between states, actions, and corresponding rewards at the moment of decision-making, the knowledge is obtained by "Exploration" and "Exploitation" strategies, similar to the learning experience of humans. Because the design process of reinforcement learning algorithms does not require an in-depth study of the background knowledge of the problem, and the training process does not require the preparation of labeled data in advance to assist the training, it is gradually favored by experts in the field of combinatorial optimization. However, reinforcement learning also has obvious limitations [40]: as the solution scale of the task planning problem increases, its training efficiency decreases steeply or even fails to converge. Different problems involve different training algorithms and strategies, and the training effect is vastly different.

From the above analysis, whether it is the mainstream mathematical planning algorithms, heuristic algorithms, and metaheuristic algorithms in the field of combinatorial optimization, or reinforcement learning, which has been in constant demand in recent years, it is difficult to independently solve the complex imaging satellite task planning problem from the perspective of the known public research results. Therefore, by reasonably decomposing the problem and designing corresponding algorithms for the characteristics of different solving processes, the advantages of different algorithms can be more effectively utilized to achieve the effect of "1 + 1 > 2." The idea of solving the imaging satellite task planning problem in a hierarchical manner is born. The experience of solving a large number of complex combinatorial optimization problems shows that decomposing the satellite task planning problem into several solution processes with logical sequence relationships and solving these problems hierarchically can effectively reduce the complexity of the satellite task planning problem, thus ensuring the quality and efficiency of the solution [41, 42, 43].

Starting from the actual application situation, this book combs through the satellite operation and control process, and designs the basic imaging satellite task planning system framework using the unified modeling language (UML), analyzes its operation flow, functional modules, data structure, etc., which provides the basis for further understanding and defining the imaging satellite task planning problem.

Based on the design of an imaging satellite task planning system, the following questions are the key scientific concerns of this research:
① How can the imaging satellite task planning problem be decomposed?
② How to model and analyze the solution process in each subproblem?
③ Can a relatively generalized method be found to ensure the quality of the solution to the imaging satellite task planning problem?

In this book, we consider the imaging satellite task planning problem as a bilevel optimization model: the upper level is the task assignment process, which decides the visible time window corresponding to the execution of each imaging task; and the lower level is the task scheduling process, which decides the specific moment of execution of each imaging task within the determined visible time window.

In this model, the task scheduling process can be modeled as a class of combinatorial optimization problems based on the results of the task assignment process. The main challenge for this problem comes from the complexity of the constraints. The variety and number of constraints are enormous, a feature that is less considered in classical combinatorial optimization problems. The study proposes an efficient constraint modeling and constraint checking method by combing the constraints in the imaging satellite task planning problem, classifying the constraints and treating each class of constraints separately. Based on this constraint checking method, two deterministic algorithms are designed to solve the task scheduling problem. Through theoretical analysis, the optimality of the two designed algorithms under specific conditions is proved; both algorithms do not contain the processing of specific constraints in the main process, which maximizes the generality and reliability of the algorithms in different engineering problems.

Relative to the task scheduling process, the task assignment process is difficult to be solved independently using the traditional optimization theory. This is because it is difficult to find an independent objective function to evaluate the goodness of the assignment process, and the quality of the assignment scheme is ultimately evaluated by the task scheduling result, which is not only related to the constraints and assignment algorithms considered in the task assignment process, but also inextricably related to the solving process of task scheduling. The empirical formula of task assignment obtained through reinforcement learning method training to guide task assignment is an effective way to improve the accuracy and rapidity of the task assignment process [44]:

(1) The task assignment process can be described as a class of sequential decision-making problems [45]: according to the specifics of the task queue and resources, the resources select appropriate tasks one by one, and update the information of the task sequence and resources after each assignment. By describing this process as a finite Markov Decision Process (MDP), a satisfactory solution can be obtained by reinforcement learning algorithm;

(2) Satellite constraints and capability parameters are varied and complex, but these conditions do not change in satellite task planning scenarios. The analysis of these complex conditions and parameters cannot be bypassed using traditional operations research algorithms, whereas in the problems that reinforcement learning is good at solving, these factors correspond to the environment part of the MDP model, and the algorithms only need to call the outputs of the environment part based on the decisions made, without discussing the internal principles [46].

(3) The optimization objective of the task assignment process is difficult to be expressed by explicit mathematical formulas, but its decision variables, objective function, and constraints are explicit. That is, after a series of complex operations, each assignment scheme has a unique objective function value corresponding to it. Therefore, it is

more advantageous for the upper-level task assignment process to be described as an MDP model than to be established as a mathematical planning model.

In summary, the research of the upper-level task assignment process focuses on how to design a reasonable and efficient MDP model for the problem, and how to design and improve the reinforcement learning algorithm in order to improve its solving efficiency in the task assignment process.

After comprehensively considering the difficulty of the imaging satellite task planning problem and profoundly analyzing the advantages and shortcomings of classical optimization algorithms and reinforcement learning methods, we study algorithms by integration of deterministic algorithms and reinforcement learning to solve the imaging satellite task planning problem, modeling the upper-level task assignment process as an MDP process and using reinforcement learning algorithms, as well as the lower-level task scheduling process as a mathematical planning process and designing deterministic algorithms to solve.

How to reasonably build a bilevel optimization model and a solution framework integrating deterministic algorithms and reinforcement learning to solve the complex imaging satellite task planning problem is the key and foundation of this book. Solving the practical problem of the satellite industry in this idea and exploring the general laws of the designed methods in solving complex problems are of great significance in promoting theoretical research in the fields of reinforcement learning and operations research, as well as engineering applications in the field of satellite task planning. The significance of the study is summarized in the following four aspects:

(1) The bilevel optimization model breaks the boundary between the conventional single imaging satellite task planning model and the multisatellite collaborative task planning model, and realizes a unified description of the single-satellite and multisatellite task planning problems by designing the task assignment process as a selection of the visible time window of the task, which not only standardizes the decision variables of the task assignment process, but also reduces the difficulty of solving the task scheduling process, which is conducive to the study of both process problems for research. Compared with the widely used mixed integer programming model, this model proposes a more reasonable scheme for the standardized description of the imaging satellite task planning problem.

(2) The deterministic algorithm for solving the task scheduling problem is discussed in depth, and the optimality of the algorithm under specific conditions is proved. Based on the proved conclusions, the imaging satellite task scheduling problem can be further simplified to provide a theoretical basis for the subsequent research work of researchers in the field of imaging satellite task planning. Furthermore, the process of proving the optimality of the algorithm given in Chapter 4 can be applied to the process of proving similar problems, providing a feasible idea for researchers of other practical problems in the field of combinatorial optimization.

(3) In the whole process of integrating the deterministic algorithm and the task planning algorithm of reinforcement learning, the decoupling of the decision-making

process and specific constraints has been realized, i. e., the specific constraints can be regarded as a black-box model, and the entire integrated algorithm's solution process does not analyze and deal with the specific constraints, but only needs to call the relevant function to read the output results during the process of checking the constraints, calculating the profit, and so on. This solution idea can maximize the generality of the algorithm and, at the same time, standardize the description of the problem. It is conducive to the unified modeling and solving of various practical problems and promotes the standardization of theoretical research in this field.

(4) For a long time, there has been a huge gap between theoretical research and practical application in the field of imaging satellite task planning: due to the need to consider too many constraints and practical factors in practical application, most theoretical research results are difficult to be applied to practical problems, or advanced algorithms cannot get a good solving effect in practical problems. In Chapter 6, the combined optimization algorithm integrating deterministic algorithm and reinforcement learning is applied to the task planning problem of "SuperView-1" to comprehensively apply all the research results in this book, which proves the effectiveness of this method in solving the practical problems, and narrows the gap between the theoretical research in the field of imaging satellite task planning and the practical application.

1.4 Features and innovations

The research work in this book starts with the actual imaging satellite operation and control process to study the generalized modeling and solution method for imaging satellite task planning problems. From shallow to deep, the research is carried out in terms of system design, problem analysis, establishment of learning bilevel optimization solution framework, mathematical model establishment and solution of task scheduling and task assignment process, and engineering application, so as to gradually reveal the scientific laws and application conclusions of the imaging satellite learning bilevel task planning technology. The innovations of this book can be summarized as follows:

(1) The imaging satellite task planning system was designed to achieve a standardized description of the workflow, collaboration, business logic, and functional structure in the operation and control process of imaging satellites, and detailed analysis and design of the system's use cases, structural objects, and behavioral objects based on the visual modeling technology common to software design. Research on the imaging satellite task planning system provides an effective paradigm for the satellite controllers, system developers, and scientific researchers to understand the problem.

(2) A bilevel optimization model for the imaging satellite task planning problem is proposed. First, the basic assumptions of the research process and definition of the imaging satellite task planning problem in this book are proposed. Then the input, output, objective function, and constraints of the problem are analyzed in detail to realize

the decomposition of the problem, and a bilevel optimization model for imaging satellite task planning problem is proposed.

(3) A solution framework integrating deterministic algorithm and reinforcement learning is proposed. The framework integrates deterministic algorithm and reinforcement learning, where the deterministic algorithm is used to solve the task scheduling problem while the reinforcement learning algorithm is used to solve the task assignment problem. Through the continuous interaction of these two algorithmic modules, training data is generated to realize the training process of the value function that is eventually used for the task assignment problem.

(4) A heuristic algorithm based on density of residual tasks and a dynamic planning algorithm based on task sequencing are proposed for the task scheduling problem. Both algorithms can theoretically prove their time efficiency and optimality, and the handling of constraints ensures the universality of the algorithms when facing different complex constraints. The experimental results show that the two algorithms have their own advantages, in which the heuristic algorithm based on residual task density has a significant advantage over the dynamic programming algorithm based on task sequencing and other metaheuristic algorithms in terms of computational efficiency, and the dynamic programming algorithm based on task sequencing prevails in the vast majority of scenarios in terms of solution accuracy.

(5) An improved deep Q-learning algorithm is proposed for the task assignment problem. First, based on the characteristics of the problem, the MDP model for the task assignment problem is established, and then the deep Q-learning algorithm is designed and pruned with domain knowledge and constraints to improve the training efficiency. From the analysis of the experimental results, the algorithm can converge within a small number of iterations and has a good generalization ability in different test data.

(6) The research results have been verified in the simulation planning scenario of "SuperView-1" commercial remote sensing constellation. By sorting out the interface and data structure of the actual engineering problems and analyzing the specific constraints, the imaging satellite task planning problems in the corresponding projects can be extracted. By applying the research results to "SuperView-1" constellation and analyzing the performance of the algorithm in daily planning scenarios, the methods proposed in this book can stably obtain a high total profit in complex real-world scenarios, proving that the learning-based bilevel task planning theory and method have a high value in engineering and application.

1.5 Content framework

This book investigates a learning-based bilevel optimization approach that integrates deterministic algorithms and reinforcement learning and its application in the field of imaging satellite task planning. The organization of the paper is shown in Figure 1.4.

Chapter 1: Introduction	Basic concept	Motivations & Significance	Features & Innovations

Chapter 2: Imaging satellites task planning system design

Requirement Analysis	Control process	Collaboration
	Business logic	Function structure

Design philosophy	Design concept
	Design principles

UML modeling	Use case
	Structured object
	Behavior object

Chapter 3: Imaging satellite task planning problem and bilevel optimization model

Problem analysis	Definition	Assumption	Component

Problem decomposition	Scheduling model	Bilevel planning model
	Assigning model	

Modeling	Mathematical programming model for task scheduling
	MDP model for task assignment
	Learning-based techniques

Chapter 4: Research on task scheduling problems of imaging satellites based on deterministic algorithms

Constraint analysis

Constraint checking algorithm

HADRT	DPTS

Simulation experiment

Chapter 5: Research on task assignment problems of imaging satellites based on reinforcement learning

The finite Markov decision model

Action	State	Reward	Value

Improved Deep Q-Learning Algorithm

Simulation experiment

Chapter 6: Application study on "SuperView-1" imaging satellites task planning

Background	Modeling & solving	Bilevel optimization model	Simulation experiment
System design		Learning-based algorithms	

Chapter 7: Summary and Outlook

Modeling · Solving · Application

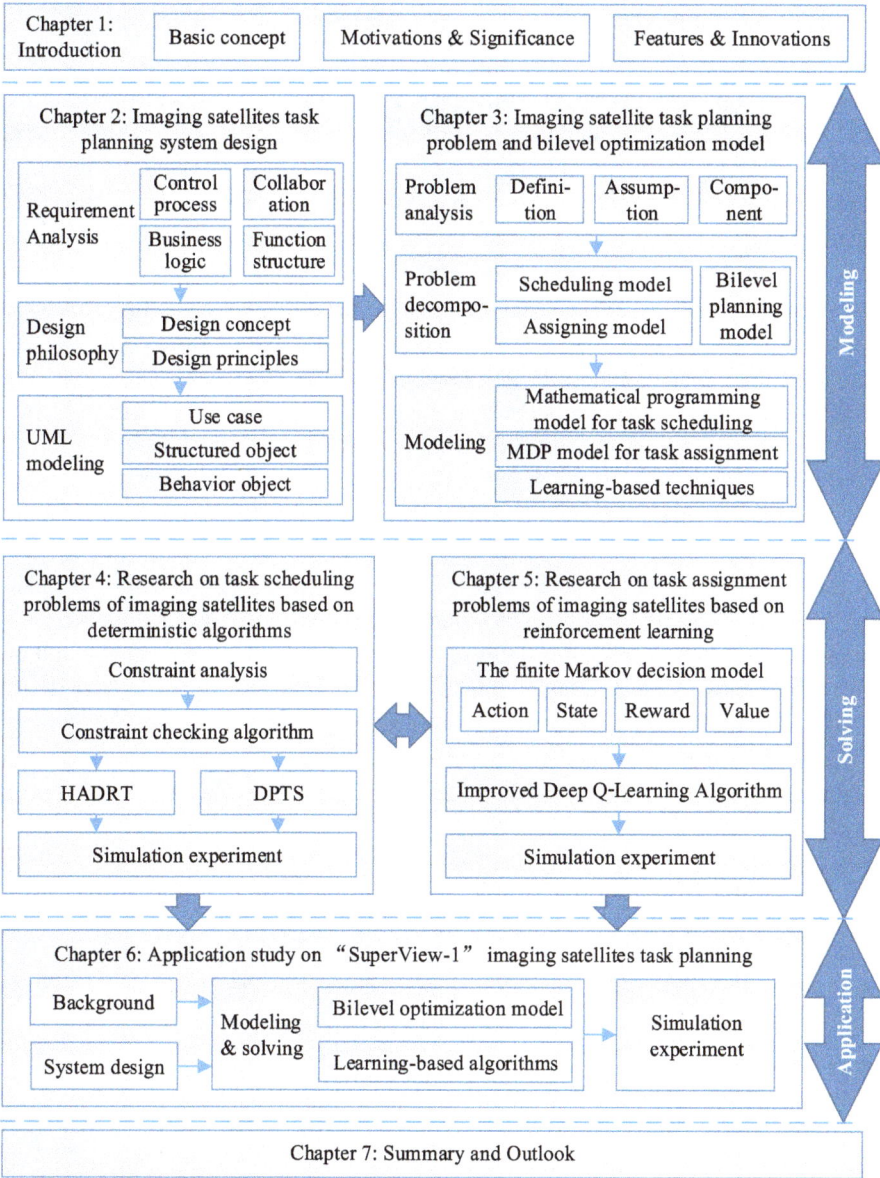

Figure 1.4: Thesis organization chart.

2 Imaging satellites task planning system design

This chapter helps beginners quickly understand the imaging satellite task planning process by systematically modeling the elements in imaging satellite task planning. Based on the analysis of the characteristics of the imaging satellite task planning system in practical engineering applications, a universal task planning system and its related functional architecture compatible with the current mainstream management and control modes and the future development trend of the satellite industry is proposed. First, the terms and their definitions related to the imaging satellite task planning process are given, and the international typical satellite task planning systems and projects are investigated; then the characteristics and functional requirements of the imaging satellite task planning system are studied from the overall situation of the four aspects of the imaging satellite: operations and control process, collaboration, business logic, and functional structure. Second, the design concept and principles of the imaging satellite task planning system are proposed by utilizing the software development technologies; a set of universal task planning system and its related functional architecture are proposed based on the analysis of the characteristics of the imaging satellite task planning system in the actual engineering applications and future development trend of the satellite industry. Finally, the system's use case model, structural objects, and behavioral objects are designed in detail by using the unified modeling language (UML), a standard visual modeling technology in software development, to deeply analyze the interrelationships of the system's internal components, the basic features, and their connotations, to lay the foundation for the analysis of imaging satellite task planning problems.

2.1 Terms definition

Because of the highly specialized nature of imaging satellite task planning, this chapter begins by defining the major terminology covered in this book.

Imaging requirements [47, 48]
Imaging requirements are the relevant information submitted by the user to the operations center in order to acquire remote sensing images. They contain two major parts, namely the target attribute set and the imaging requirement attribute set. The target attribute set is to determine the geographic location of the requirement, of which the latitude and longitude of the target point are the most basic attributes; the imaging requirement attribute set is to determine the user's requirements for the image product, such as the requirements of imaging quality, imaging time, imaging angle, observation mode, and even imaging profit.

https://doi.org/10.1515/9783111585109-002

Imaging task [49, 50]

The imaging task is a collection of data, which is one of the main inputs in the task planning. It contains the corresponding set of visible time windows, the profit of completing this task, and other necessary attributes. Among them, the visible time window is a time interval that is finally obtained by a comprehensive calculation based on the target attributes combined with orbital elements, satellite payload capacity, and other information. It is cropped according to imaging requirements. This time interval is defined as the "time window."

Imaging metatask

A metatask is the basic unit of the decision object in this problem. The properties of a metatask contain attributes such as time window, imaging duration, etc., which correspond to the properties necessary to describe the individual imaging opportunities of an imaging task. Due to the periodicity of the satellite orbit, there are multiple imaging opportunities for a single task, so it also corresponds to multiple metatasks. Since it corresponds to a single imaging opportunity, a metatask contains only a single imaging window, and what needs to be decided is the actual imaging start time in the imaging window.

Orbital elements

Orbital elements are parameters used to describe a satellite's position at any time. When this set of parameters is determined, the orbit of the satellite can be uniquely determined. In the field of satellite operations, there are two ways to uniquely draw the orbit of an artificial Earth satellite: one is the Kepler orbital parameters (semi-major axis a, eccentricity e, orbital inclination i, ascending intersection equidistant Ω, perigee angle ω, and flat perigee angle M), and the second is the Two Line Elements (TLE). Among them, the Kepler orbital parameters are succinct and usually used in simulation experiments; the TLE records more data and is usually used in practical engineering projects. The simulation experiments carried out in Chapter 4 and Chapter 5 of this book use Kepler orbital parameters to portray satellite orbits in order to simplify the relevant descriptions, while the application instances in Chapter 6 apply the internationally recognized TLE two-line orbital elements to describe satellite orbital information.

Orbiting cycles

Imaging satellites fly in specific orbits as they rotate around the Earth. With the exception of the geostationary orbit, all satellite flight orbits are periodic. The interval consisting of two consecutive passes of a satellite in the same direction (ascending or descending orbit) at the same latitude is called an orbiting cycle.

Task pretreatment [29, 51]

In a broad sense, task pretreatment includes all the processes of various types of satellites in transforming from user requirements to imaging tasks, including area target decomposition, point target synthesis, etc. The task pretreatment process covered in this book mainly refers to the structured and normalized description of imaging requirements combined with the capabilities and constraints of satellites, which are processed through a series of computations into a collection of tasks required by the task planning model for subsequent computations. Most of the processes of task pretreatment are numerical computations oriented to specific inputs, and these computational processes, combined with the industry standards and requirements of the satellite industry, have been developed and precipitated over a long period of time, and have resulted in a set of mature computational methods and processes [30].

Satellite work scheme

Satellite work scheme refers to the task plan that the satellite is scheduled to perform over a period of time, as well as the configuration scheme for the imaging parameters of each task. The process of configuring the imaging parameters of a task includes both the invocation of inputs such as payload requirements, operating modes, etc., and the decision-making on the start and end times of the task. Usually, a satellite corresponds to a single work scheme in a planning period, the task planning scheme for a single satellite is interconvertible with the satellite work scheme, and the task planning scheme must ensure that all the constraints in the task planning model are met.

Satellite control instruction

Satellite control instruction is a set of command codes designed to consider fully all characteristics of satellite platforms and payloads, operating modes, and the environment. At present, the vast majority of imaging satellites are controlled by control instruction, which are usually formed as a package of codes. The execution of each action contains a string of instructions to ensure the accuracy of the action execution. A task (including imaging task, data transmission task, etc.) is, in turn, composed of a series of actions, so the execution of each task corresponds to a group of instructions. The process of generating instructions according to the task planning scheme is called the "instruction compile," and the process of translating the satellite control instructions into the task planning scheme is called "instruction decompile."

2.2 Status of imaging satellites task planning systems

A number of international institutions in the field of aerospace are stepping up their research on satellite task planning and scheduling systems to support the efficient and stable operation of spacecraft and to meet increasingly complex user requirements in a variety of different application scenarios. Typical imaging satellite task planning projects and systems available through public channels include the following: Automated Scheduling and Planning Environment (ASPEN) for the Earth Observing-1 (EO-1) satellite of the National Aeronautics and Space Administration (NASA) from the United States of America, and NASA's on-board scheduling planning and replanning project, Continuous Activity Scheduling Planning Execution and Replanning (CASPER), the Verification of Autonomous Mission Planning On-board a Spacecraft (VAMOS) of the German Aerospace Center (DRL), the Project for On-Board Autonomy (PROBA) of the European Space Agency (ESA), and the Autonomy Generic Architecture—Test and Application (AGATA) from the French National Centre for Space Studies (CNES).

(1) NASA's Automated Scheduling and Planning Environment (ASPEN)
ASPEN establishes a unified description language for complex imaging targets, and through a series of optimization calculations, uniformly generates various payload schemes for EO-1 satellites and then codes satellite instructions. ASPEN has embedded a modeling language dedicated to spacecraft task planning, which is capable of standardizing the descriptions of on-board activities, resources, states, and parameters. For the EO-1 satellite-specific modeling (i. e., modeling EO-1 with ASPEN's embedded language), ASPEN mainly models the imaging activities of the payload carried by EO-1. In terms of modeling different imaging tasks, in addition to the information provided by the regular user requirements, ASPEN considers that the priority of the imaging tasks depends on the following aspects: cloud coverage in the imaging region, solar altitude angle in the imaging region, parameters of the data transmission window prior to the overflow of the memory, the proximity of the imaging products corresponding to the tasks to those of other satellites, and the significance of the imaging region, among others.

(2) NASA's on-board scheduling planning and replanning module (CASPER) [52]
CASPER is a streamlined version of ASPEN that accepts instructions based on the goal state and then schedules a sequence of actions to reach the goal state without violating constraints autonomously. At each planning period, CASPER maintains four parts of information: the current state, the current goal, the current scheme, and the estimation of the future state (based on the results of simulation calculations of the system operation and the expected state based on the first three parts of information.) CASPER has two mechanisms for timeline advancement: fixed-time-triggered-based task planning and event-triggered-based task planning. The event-triggered task planning is based on the occurrence of the event to determine the time point of the state update. According to feedback data from the hardware sensors or systems, the current state and the target state are updated. Then CASPER iteratively optimizes the task planning scheme and creates a new scheme for the future period of the satellite task planning. The system structure of CASPER and the logical relationship between the functional modules are shown in Figure 2.1.

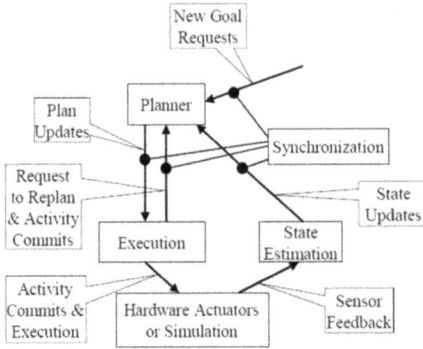

Figure 2.1: CASPER's architecture [52].

(3) DLR's verification of autonomous mission planning onboard a spacecraft (VAMOS) [53]

The main purpose of VAMOS project is to verify the capability of on-board autonomous checking and adjustment of states, in order to improve the efficiency of satellite planning and resource utilization in practice. VAMOS is tested and verified in orbit with the Fire Bispectral Infrared Detection (FireBIRD) satellite program. The architecture of the FireBIRD system is shown in Figure 2.2. The experiment consists of two main parts: the on-board and ground components. Ground components are located at the German Space Operations Center (DSOC), where the Mission Operations Segment (MOS) is one of its key units. Functions of the MOS consist of organizing the users' requirements, predicting resource and environment states, planning tasks by using heuristic algorithms, and uploading schemes to corresponding satellites. The main functions of the on-board component are accomplished in the on-board real-time operating system (RTOS), which manages the imaging results, environment, and resource changes in real-time employing the On-board Event Triggered Timeline Extension (OBETTE), which allows real-time adjustments of the actual resource usage on the satellite to ensure the on-board VAMOS can also quickly plan real-time tasks. By further identifying, locating, and tracking the detection in the onboard image processing and recognition module, the function of on-board autonomous discovery of abnormal events is achieved.

(4) ESA's project for on-board autonomy (PROBA)

The satellite served by PROBA is designed with three payloads, including a main payload and two additional payloads. The main payload is an observation payload with autonomy, and the user can submit target-oriented imaging requirements to the satellite (including necessary information such as the geographical coordinates of the target and the imaging satellite's imaging duration). After receiving these task requirements, the satellite can make autonomous decisions on-board according to the actual constraints, then complete the relevant actions (such as task pretreatment, payload attitude maneuvering, camera switching, taking images, reading and writing data, etc.) for corresponding payloads or equipment of the satellite, to ensure that the satellite can autonomously

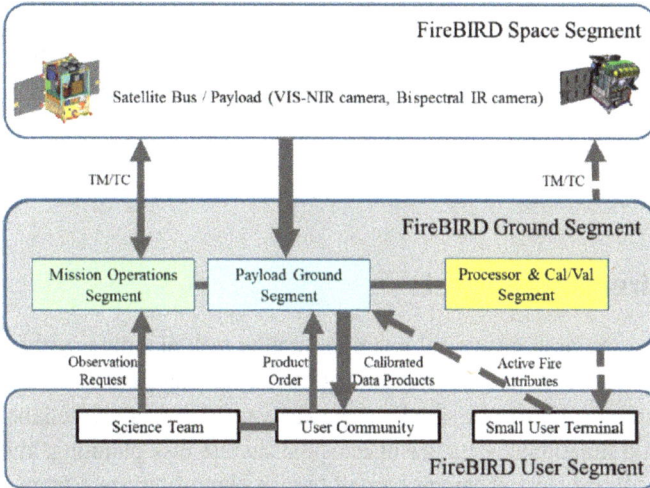

Figure 2.2: Functional structure of FireBIRD program [53].

complete the various functions such as imaging, data transmission, and data processing and analysis. In order to ensure the quality and reliability of the planning results, it is necessary to embed constraint checking, constraint optimization, and other functional modules in the on-board computer, taking into account the constraints, resources, requirements, and profits related to each task, and improving the quality of the solution as much as possible. The task planning done by PROBA on-board is relatively a short period of planning, which is designed to get a feasible task planning scheme by using as little on-board computing resources as possible.

(5) CNES's test and applications on autonomy generic architecture (AGATA) [54]

AGATA is a project of French National Centre for Space Studies (CNES), whose main research areas include space architecture design, autonomous task planning, and software upgrades and on-orbit validation. The objectives of the program are to develop a prototype system to assess the validity of advanced satellite control concepts, to develop spacecraft with high intelligence, and to verify the impact of intelligence on satellite managers. The demonstration and validation platform consists of two main segments: the space and ground segments. The space segment consists of one or several spacecraft, while the ground segment includes ground stations, operations centers, control centers, etc. AGATA's generic architecture aims to design a novel decision-making mechanism using on-board software. Its system is built on the basis of generalized modules, and each module is responsible for a part of the functions related to satellite control and operations. The AGATA task planning trigger mechanism can be described as Figure 2.3.

Figure 2.3: Trigger mechanism of AGATA [54].

2.3 Software requirements analysis

On the basis of extensive research on typical imaging satellite task planning systems and projects, combined with the national conditions of China, starting from designing the overall operations process of imaging satellites, we analyze the workflow, collaboration, business logic, and functional structure of imaging satellite task planning, and summarizes the basic requirements of imaging satellite task planning system from a holistic perspective, so as to realize the overall design of the system.

2.3.1 Operations and control process analysis

Imaging satellite task planning is the center and core of the satellite operations process (hereafter referred to as the "operations process") [55]. Therefore, in order to better understand the nature of the imaging satellite task planning problem, it is necessary to understand and analyze the operations process of the satellite first. The conceptual diagram of the operation control process is shown in Figure 2.4.

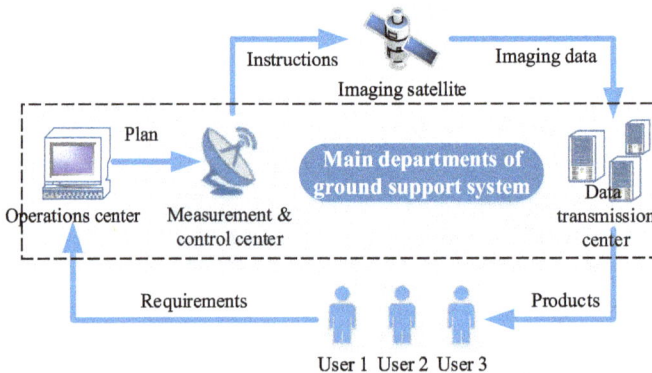

Figure 2.4: Conceptual diagram of imaging satellite operation control.

The operations process of imaging satellites is a complex system engineering that requires the collaboration of many departments. Figure 2.4 only describes the simplified workflow of the operations process of the imaging satellite, and it is briefly described as follows: the operations center receives imaging requirements from users. After collecting and organizing these imaging requirements, the operations center decides which requirements can be satisfied and when and in what way, taking into account the imaging requirements and preferences of each user's requirements, the satellite's payload capacity, and platform usage constraints. Then the operations center breaks down the decision-making results into satellite work schemes, data transmission schemes, measurement and control schemes, etc., and at the same time generates corresponding control instructions according to the work scheme of each satellite. Among them, the control instructions and the measurement and control scheme are sent to the measurement and control center. Then the measurement and control center uploads the instructions to satellites through uplinks of the corresponding satellite, and the imaging satellites take action and collect the image data according to these instructions; the data transmission scheme is sent to the data transmission center, which cooperates with the completion of the data transmission process [56]. Finally, the data transmission center receives the satellite data and processes it into image products for sending to users [57].

The operations center, measurement and control center, and data transmission center, as the crucial working departments of the imaging satellite operations and control, are all indispensable to the imaging satellite operation control process. Specifically:

(1) The operations center can be regarded as the information hub of satellite operations, which gathers all data related to satellite task planning from users, measurement and control center, data transmission centers, and imaging satellites. Through the collection, collation, calculation, and distribution of the information, a specific imaging satellite work scheme (including an imaging scheme and a data transmission scheme, etc.) can be formulated from the operations center. Based on these schemes, satellite control instructions are formed by the corresponding coding specification, and data reception plans of the data transmission stations are generated by the data transmission scheme. Control instructions and data reception plans are distributed to the measurement and control center and the data transmission center, respectively, thus realizing efficient satellite control.

(2) The measurement and control center can be regarded as the direct controlling party of the satellite. The core duties in the center of telemetry and remote control imaging satellites include two aspects: telemetry is to track and measure the satellite, to receive and process the telemetry parameters of the satellite, to calculate and update the satellite's orbital elements, satellite attitude, payload states, and other attributes about onboard working, and then to monitor the satellite to ensure that the various subsystems of the satellite are normal; remote control is mainly to transmit real-time or program control instructions to satellites through the affiliated measurement and control uplinks to realize the control of the satellite's execution of actions.

(3) The data transmission center can be regarded as the receiver of satellite data, and the process of data transmission is realized through the cooperation between stations and satellites. Based on the raw data collected by the satellite, combined with im-

age processing technology and user requirements, the data transmission center further produces the image products required by the users and distributes them in time.

The efficient work of the Earth observation satellite requires the cooperation of four parts: the operations center, the measurement and control center, the data transmission center, and the on-orbit satellites. It is often customary to refer to the operations center, measurement and control center, and data transmission center as the ground support system, while the imaging satellite or constellation as the in-orbit operation system, where each satellite contains at least a housekeeping subsystem, an antenna subsystem, an attitude control subsystem, an imaging payload subsystem, a data storage subsystem, a power subsystem, and a temperature control subsystem, etc. [54, 58]. The basic functional statistics of each component of an in-orbit satellite are shown in Table 2.1.

Table 2.1: Basic functional components of the satellite system.

Subsystem	Function
Housekeeping subsystem	Satellite or constellation operation management, control other subsystems
Antenna subsystems	Data and command reception, intersatellite or satellite-ground data transmission
Attitude control subsystem	Satellite position monitor and control, satellite attitude control
Imaging payload subsystem	Imaging parameter adjustment, execution of imaging related actions
Data storage subsystem	Storage and management of imaging and satellite telemetry data
Anomaly monitoring subsystem	Satellite working conditions monitoring
Power subsystems	Provision of satellite operating energy, charging and discharging process control
...	...

Based on the above description, it is not difficult to find that the imaging satellite task planning process is mainly done in the operations center, which determines:
① What requirements should be met?
② What resources are applied to satisfy the users' needs?
③ When is the corresponding imaging tasks executed?
④ Which uplink is chosen to upload the satellite instructions?
⑤ When are the satellite commands uploaded?
⑥ Which ground station is used to transmit the satellite data?
⑦ When is the satellite data transmitted?

To answer these questions, the operations center must collect the necessary information from users and departments related to satellite task planning and use advanced combinatorial optimization algorithms to achieve rational resource assignment and efficient task decision-making.

2.3.2 Analysis of collaboration

The collaboration diagram of imaging satellite operations is shown in Figure 2.5, where the arrows indicate the direction of information transfer between the various components of the system, and the numerical numbering indicates the logical order of operations and information transfer during satellite operation.

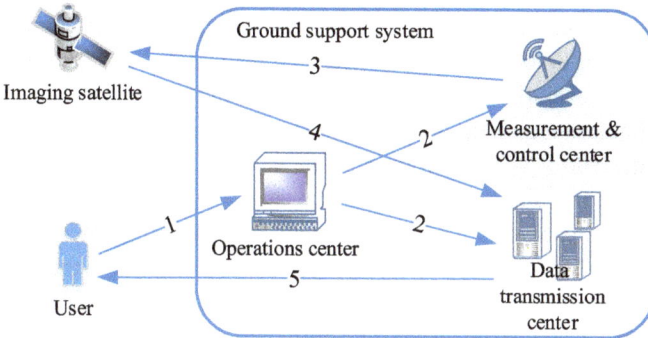

Figure 2.5: Collaboration diagram.

The user sends the satellite imaging requirements to the satellite operations center, and the requirement information includes: ① the coordinate range of the imaging area, ② the timeliness requirement, ③ the task priority, ④ the imaging mode requirement, and other specific parameters and requirements for imaging products. The operations center is responsible for organizing user requirements. Then the decomposition and merging of user requirements are realized by combining the conditions of imaging satellite orbital elements, satellite attitude maneuvering capability, satellite payload capability, the geographical environment about these requirements, etc., and the user requirements are transformed into standardized descriptions of imaging tasks. Then a planning scheme is formed from a set of imaging tasks, which should meet the constraints of imaging satellite task planning. Finally, based on the scheme, work plans of the corresponding satellite, measurement and control station, and data transmission station are generated and then distributed these plans to the corresponding departments. Satellite instructions are generated synchronously, which are distributed to the measurement and control center. These departments receive work plans and take the corresponding actions [59, 60] at a specific moment. The antenna subsystem on-board receives the instructions and forwards them to the housekeeping subsystem. The housekeeping subsystem analyzes it immediately after receiving the instruction. At the same time, the housekeeping subsystem combines all the information about the satellite and activates other subsystems to execute the actions in a specific moment, such as the attitude control subsystem adjusts the attitude of the satellite, the imaging payload subsystem executes the imaging-related actions, and the data storage subsystem records or transmits the data, etc. After the execution of the corresponding action, each subsystem refreshes its situation data to the housekeeping subsystem for saving, forming log data, and storing it in the data storage subsystem for keeping managers informed about satellite operations [61]. The housekeeping subsystem cyclically checks the the instructions and calls the corresponding subsystem until all commands have been executed.

Due to the limitations of the flight orbit of imaging satellites, it is difficult for satellites to remain visible at all times to ground facilities such as measurement and control stations and data transmission stations. The relevant calculation module in the current imaging satellite task planning system mainly simulates the satellite flight process and calculates a series of state information, and comprehensively applies physical models, mathematical models, simulation models, and other means to construct the auxiliary calculation functions required for task planning, such as the orbit calculation model, ephemeris forecast model, Earth shadow forecast model, station forecast model, attitude maneuver time calculation model, memory consumption model, and power consumption model. These models provide the necessary information to support the task planning process, and their calculation accuracy is closely related to the goodness of the task planning scheme. This book is not concerned with optimizing the computational accuracy and efficiency of the auxiliary models, such as the orbit forecast model, memory consumption model, etc., but focuses on how to improve the quality of the task planning scheme and the computational efficiency of the task planning process under the condition that the necessary information and auxiliary models are determined and known.

2.3.3 Analysis of business logic

Based on the analysis of the collaborative relationship, it can be obtained that the imaging satellite task planning process and the task execution process are carried out alternately on a rolling basis, i. e., the instructions executed by the on-board system are determined based on the task planning program of the previous cycle; the ground system activates the satellite task planing procedures to form the satellites' work of the next cycle. According to the urgency of the user's request, the importance of the task, and the total number of tasks, the task planning cycle of imaging satellites is generally set to vary from one day to one week. The Gantt chart of the components of the imaging satellite task planning process is shown in Figure 2.6.

Processes of task planning, data transmission task planning, and instructions generating have to be completed and sent to the corresponding subsystems before the end of the previous planning cycle. The operations center collects the users' requirements, standardizes these tasks for planning, and generates satellite instructions and data transmission plans, and then sends them to the measurement and control center and the data transmission center, respectively. If a new task requirement arrives after this, it is necessary to analyze whether a time window of available uplinks [62, 63, 64] exists

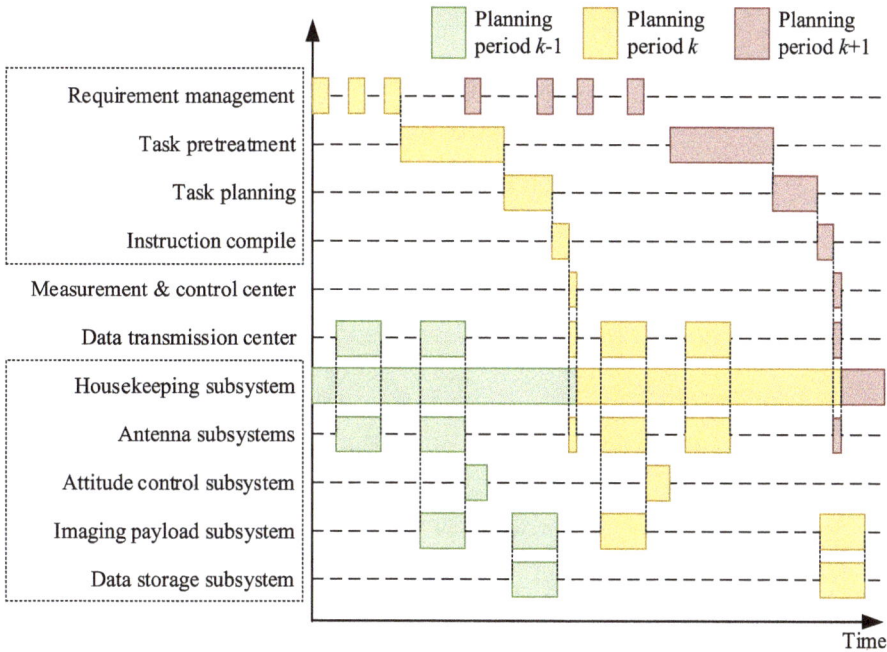

Figure 2.6: Gantt chart of imaging satellite task planning process.

in the planning cycle. If the task requires execution in the current cycle, but there is no uplink from the time the operations center receives the task until the latest execution time of the task, it can be directly determined that the task cannot be fulfilled; if the task is asked to execute in the next cycle, the operations center can replan the task with other unfinished tasks and override the original work scheme.

2.3.4 Design of functional structure

As the complexity of the functions of imaging satellites task planning as well as the number of users' requirements increase continuously, the efficiency and reliability of the satellite task planning process that relies solely on personnel are decreasing dramatically, and task planning systems have emerged. Satellite task planning systems are usually deployed in the operations center. They are responsible for the whole process, from collecting user requirements to generating measurement and control plans, data transmission plans, and satellite instructions. The essential components of a typical imaging satellite task planning system are shown in Figure 2.7.

Figure 2.7 shows the basic functional components of the imaging satellite task planning system and their interrelationships. The imaging satellite task planning system mainly includes four parts: resource management module, task pretreatment module,

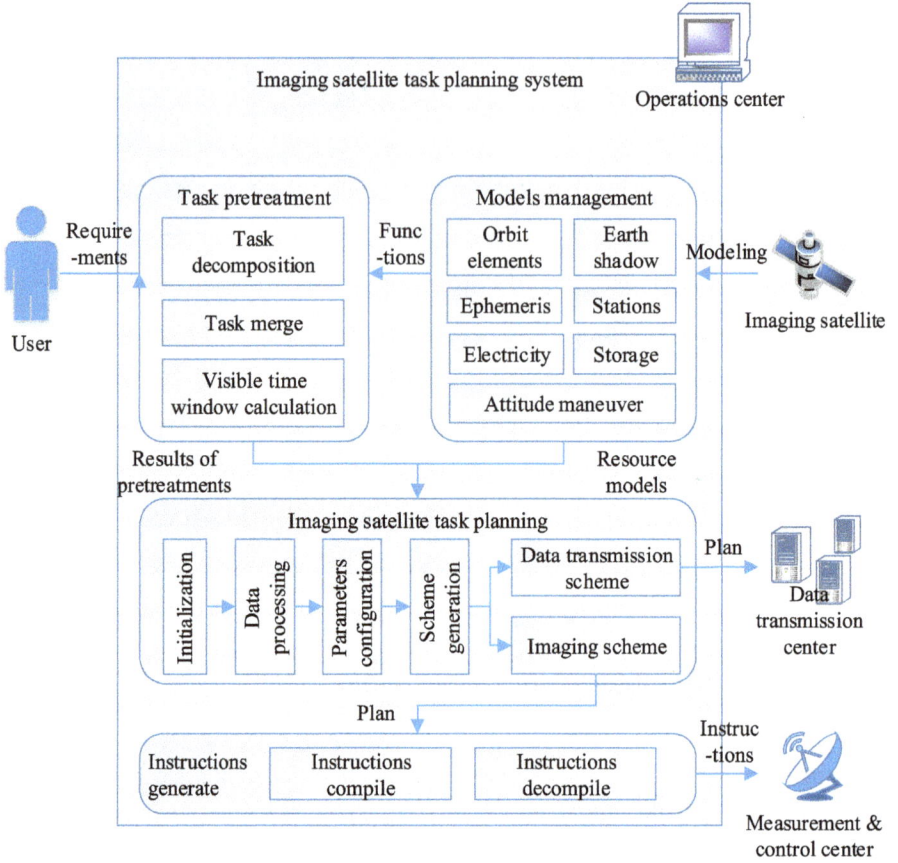

Figure 2.7: Basic architecture of a ground-based centralized task planning system.

task planning module, and instructions generation module. The visible time window of the corresponding task can be calculated by combining the geographic location of the target of the imaging satellite requirements with the satellite orbital elements, ephemeris forecast algorithms, Earth shadow forecast algorithms, etc.; the available time windows of uplinks and data transmission can be calculated by combining the measurement and control station as well as the location of the data transmission station, with the relevant information of the satellite; the task merging operation of part of the imaging satellite requirements can be realized by combining the attributes of the time windows and the geographic location of tasks requirements, which is an important part of task pretreatment [65]. Considering all the information on tasks and ground stations, a complete task planning scheme is formed by considering the objective function and constraints designed based on real-world applications. Finally, the satellite measurement and control plan, data transmission plan, and satellite instructions are formed by the corresponding modules based on the satellite work plan.

In the above process, the forecasting process of various types of resources and the task pretreatment process are the prerequisite and foundation of task planning, which are usually based on objective physical laws, hardware structure characteristics or operation, and management modes, etc., to establish simulation models or mathematical computation models to estimate the changes in the environment and state of the satellite when it is actually in operation. Because of the complex physical processes involved and the high requirements for computational accuracy, the resource forecasting and task pretreatment processes require relatively long computation time. In the process of solving the task planning problem, it is necessary to constantly call the relevant computational functions to assist the constraint checking and optimization adjustment process, so the way to improve the solution quality by using large-scale iterative strategy is not suitable for solving the imaging satellite task planning problem, which is one of the difficulties of the task planning problem compared with the classical combinatorial optimization problem. The last part is the instructions generation module, which is a process that splits the task planning scheme into corresponding work plans or instructions according to the entity that accomplishes the task. The related technology has been relatively mature and stable, so it is not discussed in this book.

2.4 System design philosophy

2.4.1 Overall design concept

At present, the vast majority of imaging satellite task planning systems in the practical applications of the satellite industry are customized, i. e., the design of auxiliary models, calculation methods, and the construction of systems for different types and models of imaging satellites for each satellite requires a large amount of manpower, material, and financial and time investment. Today, when the number of satellites in China is skyrocketing, the development of imaging satellite task planning systems by customized models has become an important constraint for the high-speed development of the satellite industry. Therefore, it is necessary to realize the standardization of some functions in the system and solidify them into universal modules based on the existing system foundation. Meanwhile, decouple the differentiated contents of different satellites (such as resource model, mathematical model, optimization algorithm, etc.) from the modules as far as possible and access to the imaging satellite task planning system in the form of plug-ins, to reduce the cost and difficulty of the system development, improve the system's versatility and efficiency, and achieve the rapid development and construction of the task planning project.

The essence of imaging satellite task planning is how to make full use of satellite resources and rationally arrange imaging tasks without violating constraints, so as to maximize the total profit of satellite use. This book designs a generalized imaging satellite task planning system as shown in Figure 2.8. The design scheme is an improved version

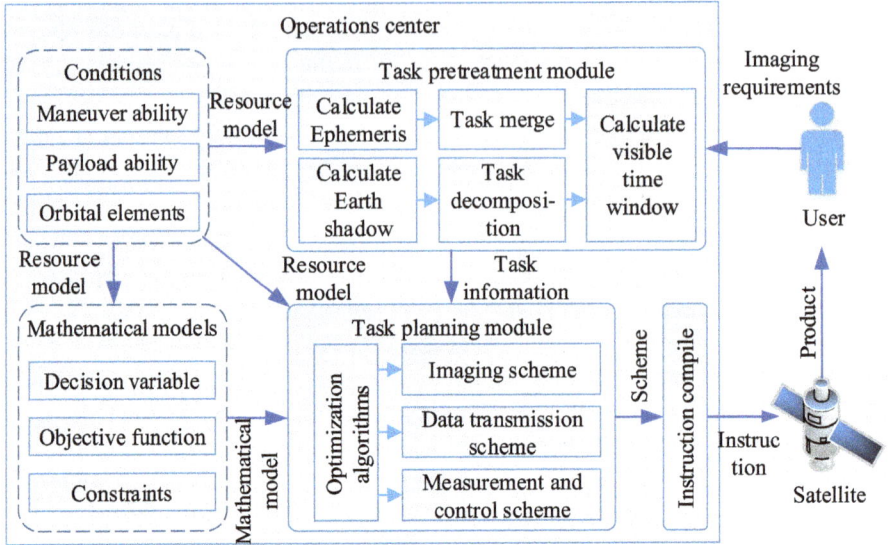

Figure 2.8: Conceptual diagram of generalized system design.

of the system architecture of Figure 2.7, which inherits the functional modules such as task pretreatment, task planning, and instructions generation from the typical imaging satellite task planning systems and divides the resource management module into two parts according to the capability parameters and computational models, and the computational models such as ephemeris forecast, Earth shadow forecast, etc. are merged with the pretreatment module, and the capability parameters are managed independently as resource parameters for different satellites. A new management module is added for the proposed satellite task planning system, which facilitates the personalized management and maintenance of decision variables, objective functions, and constraints of the task planning process, which enables the rapid adjustments of the decision preferences, environment conditions, and application objectives according to the actual needs. It can realize rapid adjustment of decision preferences, environmental conditions, and application goals according to actual needs.

This chapter aims to design a generalized and efficient imaging satellite task planning system. Through the analysis and demonstration of the whole process of imaging satellite operation and control, the current imaging satellite task planning system is improved, a novel architecture, functional composition, data structure, and logical relationship of imaging satellite task planning system are proposed, the relationship between the task planning system and other systems is analyzed, and the operations mechanism of the system is sorted out, so as to enhance the satellite task planning process, reduce the cost and complexity of the system developed by the satellite industry, improve the efficiency and flexibility of the satellite task planning process, and further explore the application effectiveness of imaging satellites.

2.4.2 System design principles

Based on China's existing imaging satellite management and control systems and the experience and achievements accumulated in long-term practice of satellite industry, facing the future enhancement of communication capability and payload performance of new remote sensing satellites, as well as the strong needs of imaging satellites in complex application scenarios such as high-efficiency information acquisition, a novel imaging satellite task planning system have been designed to improve the versatility, flexibility, and plug-in capability of the system. According to the new needs and characteristics of imaging satellite task planning, combined with the expectation of satellite management and control departments to improve the effect of satellite applications continuously, the design of the new generation of imaging satellite task planning system should follow the following principles [8, 66, 67]:

(1) Integrate processes about task planning to improve system efficiency

Looking toward the future, users have increasingly high expectations for the timeliness of the task planning process for imaging satellites; at the same time, in the face of natural disasters, environmental events, security accidents, and other emergency affairs, which are usually characterized by a greater degree of uncertainty and suddenness in both time and space. They are usually of a higher degree of importance than conventional tasks, so the system is required to provide a rapid response. The task planning process for imaging satellites involves many departments and a complex business process. In order to ensure the high efficiency of task planning and operations, it is necessary to integrate the task planning process to reduce nonessential calculations and manual intervention processes, improve the efficiency of information processing and transmission, and optimize the overall process.

(2) Reorganize the task planning function to strive for "high cohesion and low coupling"

The imaging satellite task planning process can only be accomplished through the operational synergies of multiple functional departments. These operations involve a large number of functional modules, strict calling conditions and complex logical relationships, so the design of the system's functional structure is a key element in the design of the imaging satellite task planning system. Among them, "high cohesion and low coupling" is one of the criteria for evaluating the design of a software system, which measures the degree of connection between the elements within the module and the complexity of the interface between the module and the module of the software system. The goal of this system design is to realize a high degree of concentration of functions within the module and minimize the connection between the modules to enhance the reusability and operability of the system, reusability and portability of the modules. The goal of this system design is to minimize the coupling between modules and grouping functions together as much as possible.

(3) Focus on intelligent computing business for fast and accurate decision-making

The imaging satellite task planning system is a class of management information system, the essence of which is the process of integrating, processing, analyzing, and calculating multivariate data (e. g., requirements parameters, satellites parameters, environment parameters, etc.) to obtain high-value information (e. g., the task planning scheme). This process can be subdivided into many specific computational processes, in which the core decision-making business of the task planning process should be emphasized, and other numerical computational processes should be designed around this so that all functional modules can serve the core decision-making process and enhance the overall application effect and computational efficiency of the system.

2.5 UML modeling

The overall design of the imaging satellite task planning system realizes the preliminary sorting out of the important objects, as well as the composition, logic, and relationship between the objects and the system are summarized. However, the more detailed use case model, system structural objects, and system behavior objects need to be designed in detail with the help of unified modeling language (UML). Focusing on the system design concept, based on the UML system modeling theory, the use case model, structural objects, and behavioral objects are analyzed and designed in this section, then the design sketch of the imaging satellite task planning system can be formed.

2.5.1 Use case modeling

Use case modeling is mainly applied to describe the role of the system with external objects. The use case modeling process consists of the following steps:

(1) Identifying participants

The first step in building a use case model is identifying the participants. According to the design of the capabilities and requirements of the imaging satellite operations, it is not difficult to find out that: the user puts forward the imaging requirements to the system, and also obtains the final imaging products from the system; the operator of the operations center completes the works of task receiving, task planning, instructions generating and distributing, and other works based on the imaging satellite task planning system deployed in the operations center to ensure the feasibility and efficiency of the task planning scheme; the operator of the measurement and control center receives and maintains the uploaded instructions and monitors the real-time status of the satellite based on the imaging satellite measurement and control system to ensure the safety and reliability of the satellite operation process; the operator of the data transmission center controls the relevant facilities based on the data receiving system to cooperate with the data transmission process of satellite operations. Through the above analysis, it is determined that the participants in the use case model are operators in the operations center, operators in the measurement and control center, operators in the data transmission center, and satellite users, as shown in Table 2.2. It is worth explaining that the operators in the operations center, measurement and control center, and data transmission center refer to all the entities that can complete the operations of the corresponding system to realize relevant functions, which can be natural persons, special automatic equipment with specific information processing and decision-making authority in the system, or program control modules. The users refer to the entity that can provide real-time task requirements and send them to the task planning system. Under the existing satellite operations systems, most of the task requirements are reported by individual users, government departments, and companies with satellite application needs, and in the task planning process for the future intelligent satellite network, imaging satellite requirements may be transmitted to each other through the communication links between satellites or generated by in-orbit satellites independently according to the actual needs.

Table 2.2: Participants in the use case model.

Participants	Possible entities
Users	Individual users, staff of government departments, demand intelligence generation module, etc.
Operators in the operations center	Staff in the center, program control modules, other automation equipment, etc.
Operators in the measurement and control center	Staff in the center, program control modules, other automatic equipment, etc.
Operators in the data transmission center	Staff in the center, program control modules, other automation equipment, etc.

(2) Determining the scope of the system

According to the design in Section 2.3.1, the operations process of imaging satellites needs to take into account at least entities including the operations center, the measurement and control center, the data transmission center, and the imaging satellites, which relate to the four systems: the task planning system, the measurement and control system, the data transmission system, and the onboard housekeeping system, respectively. There is a close intrinsic connection between the operations of these systems, and they work together to complete satellite control-related operations. The set of functions in each system that are directly related to task planning are the contents studied in this book, and the system boundaries and interfaces are determined according to the system functions and elements. Mutual cooperation based on the above four systems by working of users, operators in the operations center, operators in the measurement and control center, and operators in the data transmission center can ultimately achieve the goal of realizing the efficient operation of the imaging satellite system.

(3) Determining the demand structure

The imaging satellite task planning system needs to consider the relevant functional modules of multiple departments, and accordingly, the use case requirements can be divided into four units. Among these, the operations center, measurement and control center, and data transmission center have developed systems to complete the corresponding functions and cooperate with other candidates in the UML model to realize the whole process of operations and control of imaging satellites. After a long period of development, the functional modules related to task planning in the system have been gradually matured and solidified. The design of the task planning system in this study follows the specifications of the current satellite operations system to ensure that the designed software system has strong compatibility and can be adapted to the current satellite control process while also meeting the needs of the future development trend of intelligent and autonomous imaging satellites and large-scale constellation control.

(4) Extract use cases

By combining all the analysis processes in this section, the use cases extracted in the imaging satellite task planning system are mainly divided into the use case requirements of the four participants. The specific requirements are organized in Table 2.3.

Table 2.3: Use case requirements.

Participants	Requirements
User	(1) Submit requests information
	(2) Receive imaging products
Operators in the operations center	(1) Receive and process information on resources, requirements, environments, etc.
	(2) Prepare task planning scenarios
	(3) Generate satellite instructions, measurement and control schemes, and data transmission schemes
	(4) Distribute satellite instructions, measurement and control schemes, and data transmission schemes
Operators in the measurement and control center	(1) Receive and view satellite instructions
	(2) Update satellite instructions according to the measurement and control schemes
	(3) Maintain log information of the satellite
Operators in the data transmission center	(1) Receive data transmission schemes
	(2) Implement data transmission schemes
	(3) Processing and distributing satellite data and products

The use cases requirements of each participant mentioned above are the basic use cases of the system. In order to better meet the above requirements, the system needs to be supplemented with a number of internal use cases to realize the interaction between subsystems. Detailed internal use case requirements are closely related to the actual workflow of each participant, hardware and software conditions, management mechanisms, etc. This part is not the focus of this book, so it will not be discussed here, and readers who are interested in this field can further refine it according to the methods introduced in the book.

(5) Use case scenario description

By summarizing the above analysis, the use case scenario is described as follows:

①   According to the user's actual application needs, the imaging requirements are presented to the operations center.

②   The operations center receives and reviews the imaging requirements, pretreats tasks, and gets the imaging tasks with standardized descriptions.

③   Comprehensively utilizing satellite resources, environment, tasks, and other information for task planning and its necessary calculations, the imaging satellite task planning scheme can be obtained.

④   Based on the task planning scheme, a satellite measurement and control scheme, a data reception scheme, and satellite control instructions are formed and distributed to the corresponding participants,

⑤   The measurement and control center receives the satellite task planning scheme and satellite control instructions, updates the information, checks the completeness, consistency and correctness of the data, and adjusts the work schedule of the measurement and control station in order to cooperate with the complete to uploading the instructions to the satellite.

⑥   The data transmission center receives the data transmission scheme and adjusts the work schedule of the data transmission state with the imaging satellite to complete the process of data transmission.

⑦   After receiving the satellite control instructions, the imaging satellite executes a series of actions corresponding to the control instructions to realize the corresponding functions. The measurement and control station and the data transmission station cooperate to jointly ensure the stability and safety of the imaging satellite's working process.

Based on the participants of the system and their use cases analysis, a use case diagram can be drawn. See Figure 2.9. In this figure, "Task planning" is the key use case that affects the effectiveness of the whole system application. The use case "Task pretreatment" provides standardized data and parameters for task planning, and the vast majority of other use cases are necessary to guarantee the completeness of the imaging satellite operations and control process.

2.5.2 Structural object design

Based on the use case analysis, the logical model of the imaging satellite task planning system can be initially established while the functional requirements needed for task

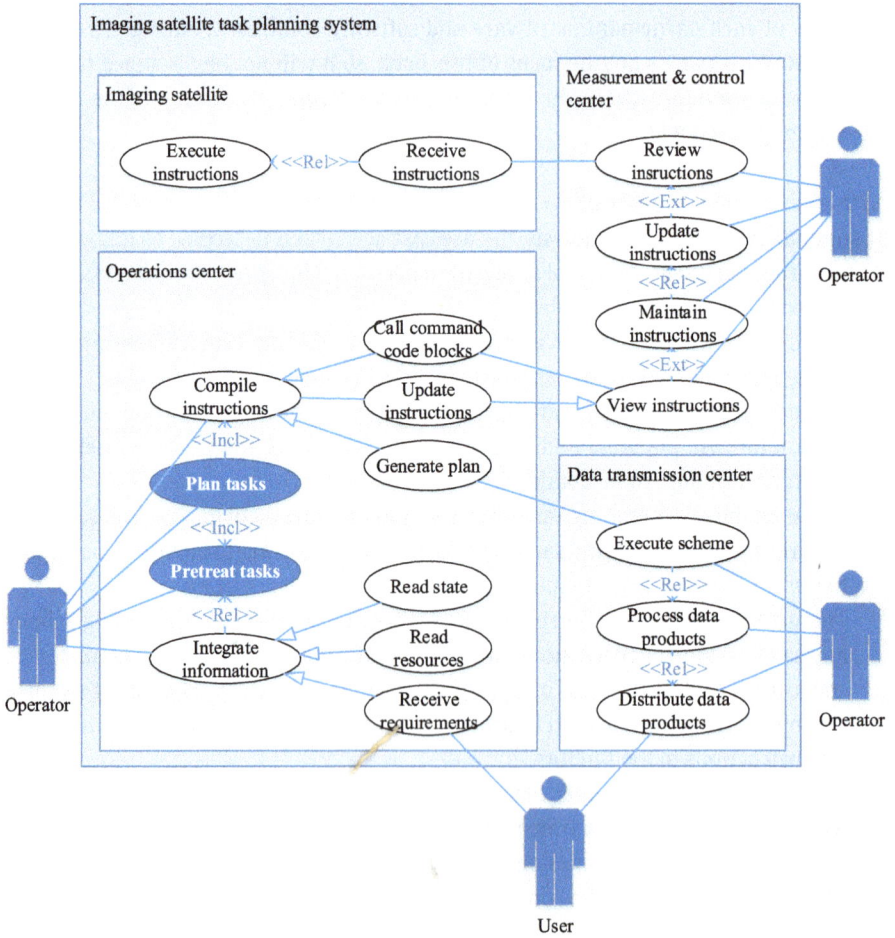

Figure 2.9: Use case diagram.

planning are sorted out. Structural object analysis is an analysis of the essential elements in the system and their interrelationships, and a structured design is carried out in order to better illustrate the storage structure of the data within the system as well as the connections between the data.

First of all, the description of each basic element in the system and its relationship is as follows: the user puts the forward requirements of the imaging satellite from the perspective of practical application, including imaging position information, imaging clarity requirements, imaging profit, and so on. By combining with the satellite resource information and auxiliary calculation function, the task pretreatment is carried out on the requirements set formed by all user requirements to realize the functions of task decomposition, task synthesis, visible time window calculation, etc. Then the imaging satellite task set is obtained. The task set contains some information describing the overall state of the tasks, which facilitates the checking of constraints during statistics and planning, as well as several pieces of task information; the task information contains information such as the task type, the task number, the task profit, the number of metatasks contained in the task, and the time parameter related to the task. A task is composed of one or more metatasks, and the attributes of each metatask contain information such as the information of visible time windows, the imaging time windows, and the duration of that metatask. The metatasks, in turn, are obtained based on the targets of requirements, and each target can link to one or more metatasks. The target information includes the target type, the country it belongs to, the coordinates of the vertex, etc. The longitude, latitude, and altitude of a point uniquely determine the coordinates. At the same time, a task set needs to be determined, which is based on the scenario information, satellite attitude, and some other auxiliary information together. The scenario information related to task planning includes satellite capability information (e. g., satellite's attitude maneuvering capability, power capacity, memory size, imaging payload's capability, etc.), satellite orbital information, and ground station information, etc.; the satellite attitude information describes the satellite's pointing information at a certain moment in time, and the other auxiliary functions, such as the calculation of attitude maneuvering time, the calculation of the task's sequencing and striping, etc., need to be determined combining with scene information, task information and target information, etc.

The next step involves analyzing the structural objects of the system and designing the static structure of the data and their interrelationships that make up the imaging satellite task planning system. Class diagrams in UML modeling provide a way of describing the static structure of the system.

(1) Identify elements and operations of class and interface attributes

Classes represent a certain set of objects with the same characteristics and properties, and interfaces represent a collection of some public operations that need to be realized through other class elements. This system mainly contains seven entity classes, i. e., requirements, task planning scenarios, tasks, metatasks, coordinate points, satellite attitudes, and auxiliary information, and two interface classes, i. e., task set interface and metatask interface. The attributes, operations, and other elements of the classes and interfaces are organized in Table 2.4.

(2) Determine the relationship between classes

Relationships between classes include compounding, aggregation, association, dependency, etc. Relationships between classes in this system are analyzed as follows. A planning scenario may not contain a task, but for the task planning process, any task needs to be processed in a specific planning set. Therefore, tasks and task sets are aggregation relationships; the planning scenario relies on information such as two functional interfaces, task sets, and satellite attitudes to realize the task planning function, so there is a

Table 2.4: Classes in the system and their attributes and operations.

Class element name	Attributes	Operations
Requirement	Target country, target type, target imaging location (target area range), etc.	–
Planning scenario	Satellite capability, Satellite orbital elements, Ground station information, etc.	Initialization, calculation of total profit, task sorting, updating of on-board resources, etc.
Task set	Number of tasks, remaining available power, remaining available storage, maximum available power, maximum available storage, etc.	Initialize task set, calculate total profit, sort tasks, calculate remaining resources, etc.
Task	Task type, task number, task profit, imaging duration, imaging time window, visible time window, number of metatasks, etc.	Initialization, calculating the amount of storage and power needed for this task, etc.
Metatask	Metatask number, imaging duration, visible time window, execution time window, etc.	–
Coordinate points	Longitude, latitude, altitude	–
Satellite attitude	Roll angle, pitch angle, yaw angle, sun altitude angle	Calculate attitude at a certain time for a certain target, etc.
Pretreatment	–	Calculate attitude maneuver time, calculate task position
Metatask	–	Perform striping and generate metatasks

dependency relationship between the planning scenario and two functional interfaces, task sets, and satellite attitudes.

Each task contains at least one metatask, and a metatask cannot form a whole on its own if the task does not exist. Similarly, an imaging request contains at least one coordinate point, and defining a coordinate point that is not associated with an imaging target can be considered meaningless for task planning. Therefore, there is a compounding relationship between metatasks and tasks, as well as between coordinate points and imaging targets.

The attributes in the satellite attitude need to be obtained through the combined calculation of the attributes in the class of planning scenario and the task, so the satellite attitude class is in a dependency relationship with various types of functional interfaces and task classes. The operation of the metatask interface needs the attributes of the requirement and the planning scenario as inputs, so the metatask interface is also dependent on the planning scenario and the requirement class.

Through the above modeling analysis, the basic data structure of the imaging satellite task planning system can be described as shown in Figure 2.10.

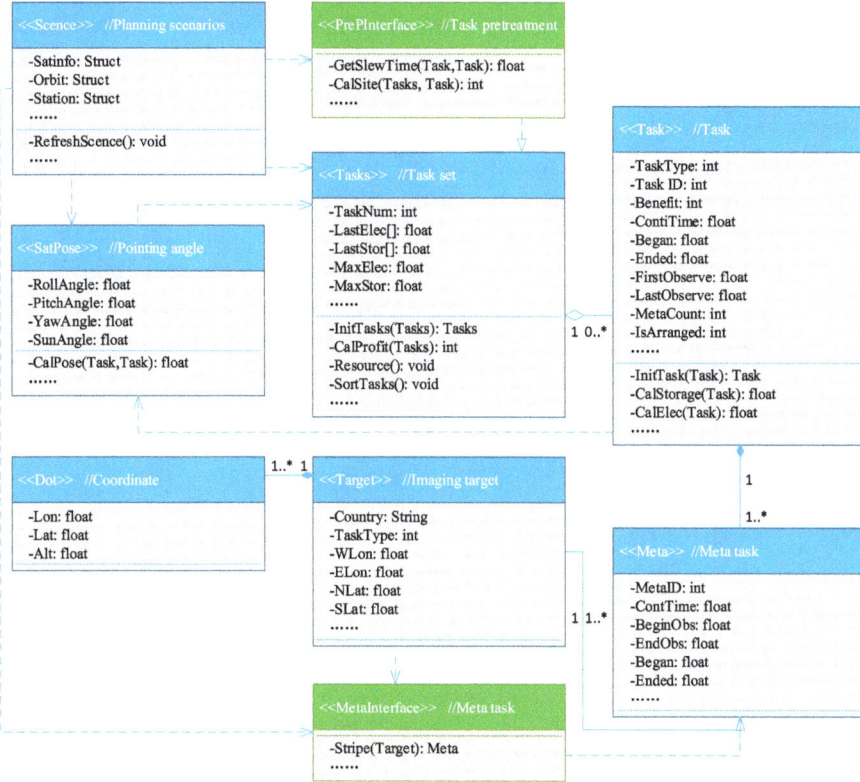

Figure 2.10: Class diagram.

2.5.3 Behavior object design

System objects have both structure and behavior. Analyzing and understanding the behavior of objects is usually represented by a sequence diagram. Sequence diagrams intuitively represent the process of the system's objects and their information transfer content and sequential relationships from the user sending the demand for the use of imaging satellites to the user receiving the final image product. This section focuses on the use of sequence diagrams in the UML model to establish object interactions arranged in chronological order and to further elucidate the implementation of use cases.

(1) Create active objects and their lifelines

Activity objects may include participants in the imaging satellite task planning system and other system objects. In the imaging satellite task planning system, the functions of the three components of the ground support part (operations center, measurement and control center, and data transmission center) are applied by the operators, and the functions of the on-orbit part (i. e., imaging satellites or constellations) are mainly realized by program modules or control instructions based on the rules set in advance. Since the specific business functions of the user, the on-orbit operation part, the measurement and control center, and the data transmission center regarding task planning are not the focus of the research content of this book, this part appropriately simplifies the behaviors of these activity objects, and focuses on highlighting the functional modules and their behavioral logics that are directly related to task planning in the operations center. Comprehensively, in the process of imaging satellite task planning, the operations center contains at least demand management, resource management, task pretreatment, task planning, command generation, and other functional modules, so the activities and logical sequence of these are discussed in this section.

(2) Create events
For each active object, messages such as Table 2.5 can be created separately.

Table 2.5: Activity list of system objects.

Object	Activities
Users	Request for imaging, receive imaging products, etc.
In-orbit operation section	Receiving commands, executing commands, data storage, status monitoring, feedback on execution of commands, etc.
Requirements management	Receive and store user requirements, standardization of requirements parameters, requirements consolidation, etc.
Resource management	Resource information management, orbital elements calculation, Earth shadow forecast, ground station forecast, auxiliary model maintenance, etc.
Task pretreatment	Task decomposition, task synthesis, task time window calculation, etc.
Task planning	Constraint handling, constraint checking, iterative optimization of scenarios, scenario generation, scenario review, etc.
Satellite instructions generation	Satellite scheduling, satellite instructions compilation, satellite instructions decompilation, distribute instructions, etc.
Measurement and control center	Receive measurement and control instructions, upload satellite control instructions, monitor satellite status, satellite in-orbit maintenance, etc.
Data transmission center	Receive data transmission schemes, execute data transmission schemes, receive satellite data, process imaging products, etc.

(3) Determine the order of activities
For the analysis of the above activities, the following summary regarding the sequence of activities can be summarized as follows:
① Users generate imaging requirements based on actual application needs and send them to the requirements management module of the operations center.
② The requirements management module receives the imaging requirements, checks the completeness of the requirements, standardizes the relevant parameters, and merges the duplicate requirements.

③ The task pretreatment module reads the standardized imaging requirements and calls the necessary information from the resource management module with the model to start the task pretreatment operation.

④ The task information obtained at the end of task pretreatment is used as an input to the task planning module, and various types of information such as resources and tasks are synthesized with the computational model to start task planning.

⑤ After the task planning is finished, the formed task planning scheme is used as the input of the satellite instruction generation module, which in turn forms the measurement schemes and data transmission schemes, compiles the satellite control instructions at the same time, and sends schemes and instructions to the corresponding objects after reviewing and approving them without errors.

⑥ After the measurement and control center receives the measurement and control schemes and satellite control instructions, it arranges the work of the subordinate measurement and control stations based on these, and realizes the upload of satellite instructions.

⑦ After receiving the data transmission schemes, the data transmission center integrates the actual operations and occupancy of the ground stations under it, and allocates corresponding facilities to cooperate with the imaging satellite to complete the data transmission activities.

Through the preliminary analysis of the activities, it can be obtained that the messages passed between the behavioral objects in this system mainly include requirements, schemes, instructions, and data products. The logical relationship between the behavioral objects of the imaging satellite task planning system can be represented by a sequence diagram such as Figure 2.11.

2.6 Summary

This chapter systematically introduces the workflow, functional composition, and business logic of the imaging satellite operations system, which can lay the foundation for readers to better understand the modeling part of the imaging satellite task planning problem in Chapter 3 of this book. Based on the current status of the satellite industry in China, facing the future direction of development, this chapter puts forward a design scheme of imaging satellite task planning system, and follows the relevant norms of software design, applying UML modeling technology to ensure the reasonableness and scientific nature of the work, which provides a feasible technical route for the satellite industry to design and develop the imaging satellite task planning system, and at the same time, provides fundamental knowledge for the relevant scientific researchers to analyze and study the imaging satellite task planning problems in the actual engineering.

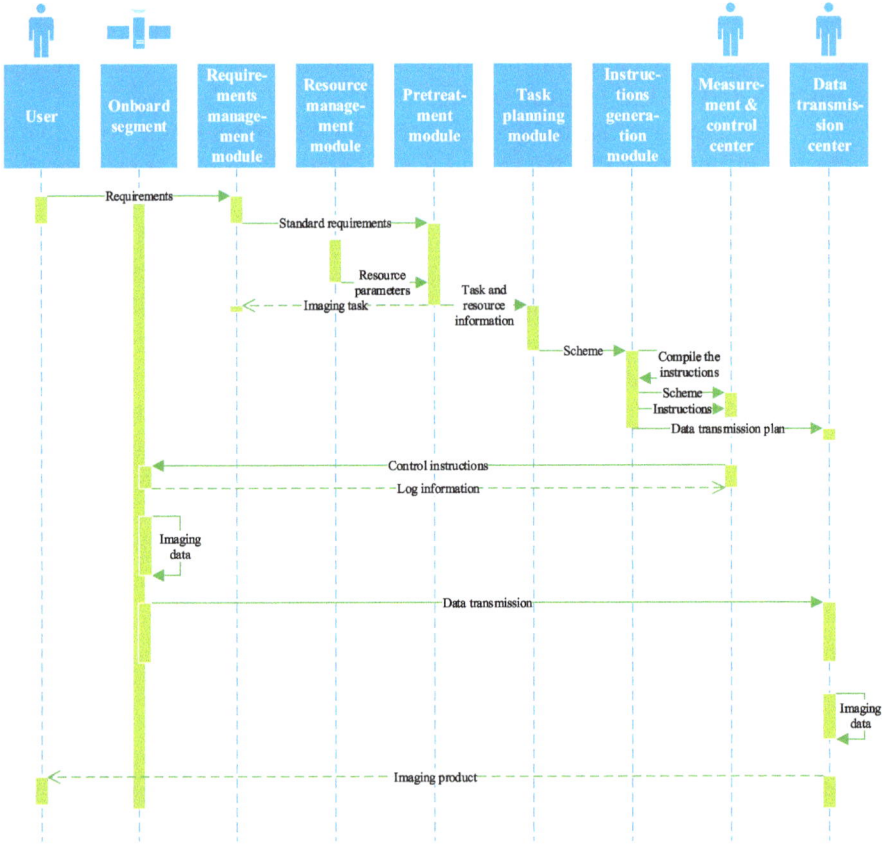

Figure 2.11: Sequence diagram.

3 Imaging satellite task planning problem analysis and bilevel optimization modeling

Based on the design of the imaging satellite task planning system in the previous chapter, this chapter first gives related definitions of the imaging satellite task planning problem. Through the investigation and analysis of the current research status of the problem, the focus of the problem under study in this book is further clarified, so as to give the basic assumptions of the imaging satellite task planning problem, define the inputs and outputs of the problem, summarize the common objective functions in the problem, and classify and discuss the constraints and their characteristics. Then a bilevel optimization model for the imaging satellite task planning problem is proposed, the functions and interrelationships of each part in the model are clarified, and the solution framework of the model is determined. Under the above model and framework, the characteristics and algorithm design principles of the upper and lower solution processes are given: the upper task assignment process is established as a finite Markov Decision Process (MDP) model and designed with a reinforcement learning algorithm, and the lower task scheduling process is established as a mathematical planning model and designed with a deterministic algorithm, so as to ensure the stability of the algorithms, the solution efficiency, and the solution accuracy in different real-world problem contexts.

3.1 Imaging satellite task planning

3.1.1 Problem definition

In terms of the complete imaging satellite operation and control process, the primary decision optimization problems are:

① What users' requests should be fulfilled?
② Which satellite is used to fulfill these requests?
③ When is the corresponding imaging task performed?
④ Through which uplink are satellite instructions uploaded?
⑤ When are the satellite instructions uploaded?
⑥ Which ground station is used for satellite data transmission?
⑦ When is the satellite data transmitted?
⑧ ...

According to the analysis and design of the imaging satellite operation control process and task planning system in Chapter 2, the main decision-making process of the imaging satellite task planning problem is completed in the operations center, which is the basis for the normal operation of the satellite and the prerequisite for satellite measurement and control centers, data transmission centers, and other related work. Therefore, the

https://doi.org/10.1515/9783111585109-003

imaging satellite task planning problem is the core problem that needs to be solved in the satellite operations center, and the quality of its solution is crucial for improving the actual operations efficiency of the satellite.

The research focus of this book is to find advanced technical methods to alleviate the contradiction between supply and demand between imaging resources and users' requirements under the condition of sufficient measurement and control resources and data transmission resources. According to the above analysis and collation, it can be obtained that the imaging satellite task planning problem starts from receiving a series of imaging tasks generated by task pretreatment, and the process of getting a satisfactory task planning scheme and satellite work schedule through the combinatorial optimization method. This book defines the imaging satellite task planning problem as follows.

Definition 3.1 (Imaging satellite task planning problem). Given n imaging tasks (an imaging task is described by a set consisting of several time windows, imaging durations, and profits), given imaging satellites, each of which has a limited payload capacity onboard, and each of which has its own corresponding utilization constraints. During the planning period, each satellite is considered for task planning for k orbiting cycles, and the task planning process is relatively independent within each orbiting cycle. Provided that all conditions and objective functions are determined and remain unchanged, a combinatorial optimization algorithm is used to compute which tasks should be selected and when they should be performed by each imaging satellite in the problem in order to satisfy all the constraints and optimize the objective function.

The schematic of the imaging satellite task planning problem is shown in Figure 3.1. It can be seen that the search space of this problem is huge. Disregarding the reduction of the solution space size by constrained merging and simplification, the search space of the problem can be estimated by the formula (3.1):

$$|\Omega| \approx \prod_{j=1}^{mk} \prod_{i=1}^{n} (we_{ij} - ws_{ij} - d_{ij}) \tag{3.1}$$

In the formula, $|\Omega|$ denotes the size of the solution space; we_{ij} and ws_{ij} denote the end time and start time of visible time window of each task, respectively; d_{ij} denotes the task imaging duration. Obviously, the size of the search space grows exponentially with the number of tasks and resources. Although in some practical problems, pruning strategies can be designed to reduce the search space according to the domain knowledge, this approach requires a large number of theoretical foundations to support as the number and complexity of constraints in the practical problems increase, while the time cost of strategy design increases, which is not conducive to the study of the problem and large-scale dissemination of the results. Therefore, in the face of the increasing scale of resources and tasks, and the increasingly complex constraints, it is imperative to establish a relatively general combinatorial optimization model.

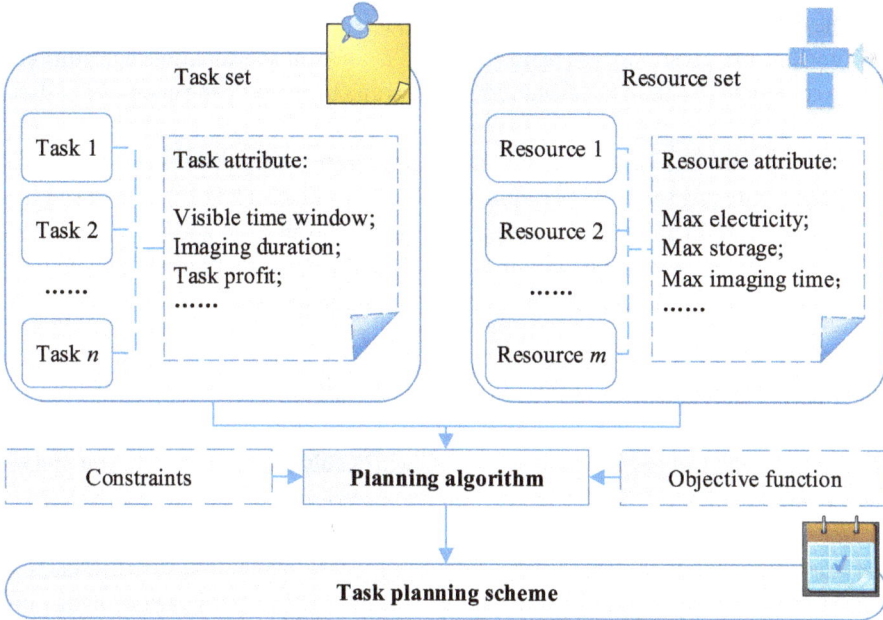

Figure 3.1: Schematic diagram of the imaging satellite task planning problem.

3.1.2 Assumptions

The model designed in this book highlights the essential features of the imaging satellite task planning problem, treats the differentiated conditions and constraints as a black-box model in a unified manner, and strives to enhance the generality and efficiency of the model, so that it can be applied to most of the imaging satellite task planning problems in practical engineering. The basic assumptions and preconditions for establishing the imaging satellite task planning problem as a learning bilevel optimization model and combining deterministic algorithms and reinforcement learning algorithms to solve it are summarized as follows:

① The number of tasks and resources is finite and known to be deterministic in advance, and the task planning algorithm will not receive new tasks and resources once it starts working (i. e., the task planning problem is considered as a class of static optimization problems).

② The objective function and constraints in the imaging satellite task planning problem can be expressed in mathematical expressions or logical expressions displays. These expressions do not change with the input conditions.

③ The impact of uncertainties such as the occurrence of resource failures on the task planning process is not considered. That is, the resource is considered to be unavailable due to uncertainties other than the availability constraints known in advance.

④ The objective function of the task planning model is a deterministic single-objective function, i. e., given two task planning schemes, an explicit evaluation algorithm can be computed to obtain the evaluation metrics and to compare the goodness of these two schemes.

⑤ Measurement and control resources and data transmission resources are always sufficient during satellite operation, i. e., the scientific problem investigated in this book focuses on mitigating the supply-demand contradiction between imaging resources and user demand without considering the optimized decision-making for the measurement and control process and data transmission process.

3.2 Literature review

The imaging satellite task planning process is inextricably linked to the efficient and reliable operation of imaging satellite systems. Through researching early work related to the operation and control of imaging satellites, it was found that there was almost no literature dedicated to the study of imaging satellite task planning before the 1990s at home and abroad. This was due to the relatively simple satellite hardware structure and single working mode at that time, and the task planning process was a simple process that could be entirely realized by hand, without the need to establish models and design algorithms to solve the problem. With the development of the satellite industry, the imaging satellite task planning problem requires more and more conditions to be considered, thus giving rise to many variants of the imaging satellite task planning problem. Starting from the basic imaging satellite task planning model, the more typical model variants and their logical relationships with each other are shown in Figure 3.2.

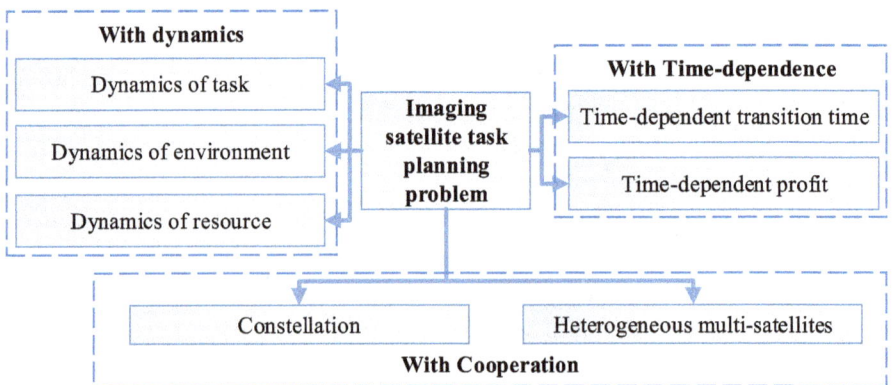

Figure 3.2: Imaging satellite scheduling problems and its variants.

The figure shows only a variant of the imaging satellite task planning problem after the change of a single condition, and specialists and scholars specialize in the imaging satellite task planning problem for different conditions. Although the imaging satellite task planning problem in real-world applications is more complex than these models, regardless of the change in the characteristics of the problem, such problems are often considered to be modeled as the following classes for research.

3.2.1 Expert system model

The expert system model emphasizes the contribution of the experience and knowledge of experts in the relevant fields to the decision-making process. In the early era of the satellite task planning problem, the problem was viewed as a "soft management" problem, i. e., the integrated management of imaging satellite tasks using knowledge of management, organizational behavior, etc., combined with the experience of experts in the field of imaging satellite applications. The problem can usually be described in the form of statements such as "the satellite should take what kind of operation in a certain kind of situation and conditions" [68]. The control of satellites is achieved by fixed rules, which in computer language is generally expressed as a fixed rule such as "IF <condition> THEN <action>." Figure 3.3 illustrates the expert-system-based task planning process for imaging satellites.

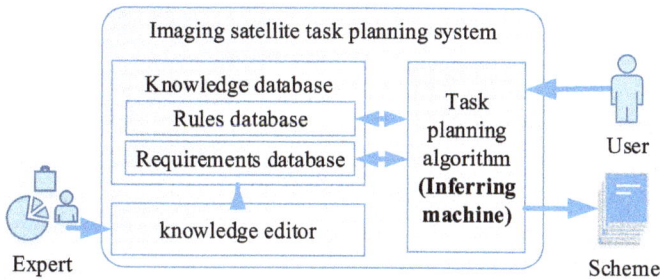

Figure 3.3: Structure of expert system-based task planning model.

In Figure 3.3, the knowledge is the core of the expert system. According to a NASA's working paper released in 1988, Barry [69] developed a scheduling and control system for Rockwell satellites based on an expert system. Expert systems can save labor costs and reduce low-level errors during manual intervention compared to manual planning, but the limitations of doing so are obvious: the rules and knowledge contain too many subjective factors of experts and managers, and it is difficult to measure the goodness of these rules from a global perspective, which does not guarantee the overall operational efficiency of the satellite. Moreover, the cost of developing and maintaining a knowledge base is often high, and with the increase in the number of satellites, expert systems that rely excessively on domain experts are gradually being eliminated in the field of satellite task planning.

With the gradual maturation of artificial intelligence technology, many artificial intelligence techniques instead of manually constructing rules are applied to the automatic construction of knowledge bases. The knowledge graph is one of the more typical representatives. However, there are still many difficulties in the application of this technology itself, such as automatic knowledge acquisition, automatic fusion of multisource information, knowledge reasoning and comprehensive application, etc., coupled with the complexity of the imaging satellite task planning problem, we have not yet found mature research results and instances of the application of knowledge mapping in the field of imaging satellite task planning in the published literature, and the application of knowledge mapping in this field needs more foundation and support.

3.2.2 General integer programming model

Integer Programming model means that the range of values of some or all variables in the model is limited to integers. Since the 1990s, the Air Force Institute of Technology in the United States has been studying the scheduling problem of the Air Force Satellite Control Network (AFSCN) through a mixed integer programming model [70]. In 2000, Wolfe and Sorensen [71] proposed the window constraint packing problem model. It designs integer planning models to describe the imaging satellite task planning problem for specific conditions in practical problems, and for more than 20 years afterward, the imaging satellite task planning problem has been customarily studied by considering integer planning models in various forms.

According to the variable composition of the constraint terms in the model, integer planning models can be categorized into linear integer planning models and nonlinear integer planning models. The basic form and components of integer planning models are the same as those of Model (1.1), with the major difference from general optimization models being that some or all of the variables in integer planning models take on integer values [72]. For the treatment of this type of models, it is typical to find a derived problem related to the satellite task planning problem, and then to find a slack problem for the corresponding problem by means of variable relaxation techniques. Based on the characteristics of the slack problem, an appropriate method is selected for solving [73].

The integer planning model can objectively reflect the nature of the imaging satellite task planning problem, but this approach to solving this problem has high requirements for the expression and processing of elements, requiring deep operational research skills and rich domain knowledge of satellite operation and control, which with high learning costs [74]. In addition, the integer planning model is less extensible and requires analysis and processing of each specific constraint in the model [75]. Due to the large number and complex form of constraints considered in the imaging satellite task planning problem, it is difficult to model and solve the imaging satellite task planning problem simply as an integer planning model for large-scale generalization and application.

3.2.3 Classical planning problem model

The imaging satellite task planning problem, as one of the typical applications in the field of combinatorial optimization, has long been of interest to scholars in the field. After decades of research, scholars have been looking for more accurate and faster methods to solve it. One of the ideas is to map the imaging satellite task planning problem to classical mathematical planning models, such as the resource-constrained project scheduling problem [76], TSPTW [45], JSP [17], etc. In addition to this, the imaging satellite task planning problem is often mapped to classical models such as the knapsack problem model [72], the directed graph model [77], and the resource assignment problem model [78]. Problems are mapped onto these classical models based on various assumptions and deformations and solved using well-established solution algorithms.

In order to minimize the total time of satellite observation, Abramson et al. [79] described the multisatellite task planning problem as a shortest path problem in graph theory and solved the problem using classical algorithms. Bai [80] established a scheduling model based on the shortest path problem for the complex multitask scheduling problem. Jun Wang and Jun Li [81] established an autonomous satellite task planning model based on a loop-free directed graph with dynamic topology for electromagnetic detection satellite payload characteristics. Jean Berger [77] approached the problem from two perspectives of global and phase, global optimization based on the satellite task planning multiple constraints to establish a multiobjective task scheduling model, and stage optimization establishes a multiobjective directed graph model for integrated scheduling. Guansheng Peng [78] abstracted the agile Earth observation satellite scheduling problem as a multiconstraint oriented problem with multiple time windows.

These classical mathematical models provide strong theoretical support for solving practical problems. However, as more and more realistic conditions and complex constraints are considered in imaging satellite task planning problems, the above models are usually oversimplified in describing real engineering problems, which leads to the difficulty of classical mathematical planning models in portraying the essential features of real engineering problems. Philippe Baptiste et al. [82] attempted to plan the constraints of scheduling problems formally, but many constraints in the actual satellite task planning process are still difficult to be described in a standardized way by a unified modeling language.

3.2.4 Constraint satisfaction problem model

Constraint Satisfaction Problem (CSP) models use the range of values of variables or the relationship between variables as constraints, and use this as a basis for finding a solution that satisfies all the constraints [83]. This class of models emphasizes on analyzing the state structure of the model by means of constraint propagation and constraint elimination, which in turn plays an important role in helping the description and simplification of large-scale multiconstraint problems. Yang Liu [84] modeled the dynamic scheduling problem of reconnaissance satellites as a class of CSP models and proposed a conceptual model of constraints. Yongming Gao [85] proposed a satellite-oriented autonomous task planning CSP model by combining the spacecraft system modeling approach and the task planning domain modeling approach. Song Liu [86] addresses the problem of autonomous planning for agile imaging satellites, and establishes a problem model based on the timeline constraint network on the basis of analyzing the main constraints. On this basis, the CSP model for the satellite task planning problem considers the visible time window constraints of the task. Ackermann et al. [87] investigated the CSP model for multiple remote sensing satellites. Beaumet [88] used the Planning Domain Definition Language (PDDL) to build a CSP model for satellite action execution, and based on this model to determine whether the satellite's action scheme satisfies the constraints, and then decide whether the corresponding action is executed or not.

In a word, different imaging satellite task planning problems can be modeled as corresponding CSP models based on specific features. These constraints are easy to describe and understand, have strong extensibility, and show good applicability. Thus, this modeling tool is widely used both in theoretical research and engineering applications in the field of imaging satellite task planning. However, this modeling approach encounters similar bottlenecks as integer planning models in solving complex, large-scale task planning problems: the efficiency of solving complex CSP models by means of backtracking or local search is low, and the difficulty of model optimization is gradually increasing.

3.3 Essential elements of the imaging satellite task planning problem

3.3.1 Inputs

The input parameters for the imaging satellite task planning problem consist of two main components: resource information and task information.

Resource information is the set of all parameters that characterize a resource. Since after the task pretreatment process, an imaging task is transformed into a collection of metatasks with a number of visible time windows (i. e., a number of observation opportunities, each of which corresponds to a metatask) and input into the model for processing, the observation opportunities of a task on multiple imaging satellites and the observation opportunities of multiple orbiting cycles on a single imaging satellite are not inherently different in terms of the form of input to the model, the difference lies in the fact that certain constraints need to be determined jointly with the task planning of other orbiting cycles. If the specific constraints are separated from the main decision-making process of the model, it is possible to realize the unified modeling of the single-satellite task planning problem and the multisatellite task planning problem.

Following this line of thought, each orbiting cycle in each satellite is considered as a separate resource: assuming that there are m satellites, each of which is considered for tasks within the planning period of k orbiting cycles, there are in total in this problem mk resources are considered uniformly. Each resource has some capacity limitations, such as satellite power limitations, storage limitations, and so on. The resource information describing the problem in this way is the basis for standardized modeling of the task planning problem for different types of imaging satellites. The mathematical description of the resource information takes the form of equation (3.2) and equation (3.3):

$$\mathbf{RS} = \bigcup_i \bigcup_j RS_i^j \tag{3.2}$$

$$\forall i, \quad RS_i^j = (C_1^j, C_2^j, \dots) \tag{3.3}$$

$$i = 1, 2, \dots, m \tag{3.4}$$

$$j = 1, 2, \dots, k \tag{3.5}$$

where **RS** denotes the set of resources in the task planning problem; the number of elements in the set is mk; each resource RS_i^j in turn needs to be given the intrinsic attributes C relevant to the task planning process, e. g., power capacity, size of the storage, and so on. These attributes are recorded and used to subsequently construct constraints for the model.

The attributes of each task TS_i can be described by a collection of triples: the visible time window $[ws_i^j, we_i^j]$, imaging duration d_i^j, and profit p_i^j. These attributes are directly related to the task number i and the resource number j. The time window, imaging duration, and even the profit may be changed for one task on different resources, because of the process of task pretreatment and users' requests. The set of all task attributes i. e. the set of this triad, the set of task information can be described by the equation (3.6) to the equation (3.8):

$$TS_i^j = \begin{cases} ([ws_i^j, we_i^j], d_i^j, p_i^j), & \text{task } i \text{ in resource } j \\ \emptyset, & \text{otherwise} \end{cases} \tag{3.6}$$

$$TS_i = \{TS_i^j \mid j = 1, 2, \dots, m\} \tag{3.7}$$

$$TS = \bigcup_{i=1,2,\dots,n} TS_i \tag{3.8}$$

3.3.2 Outputs

The resources corresponding to each task and the task start time are usually used as decision variables in the imaging satellite task planning problem, and are represented in this model by r_i and es_i, respectively.

r_i denotes the resource number corresponding to the execution of task i. If a task is not planned, then the value of r_i is set to −1 (see equation (3.9)):

$$r_i = \begin{cases} -1, & \text{task } i \text{ is unplanned} \\ j, & \text{task } i \text{ to be executed on resource } j \end{cases} \tag{3.9}$$

es_i and ee_i denote the start and end time of the execution of the task i. These two variables are not assigned values when and only when $r_i = -1$, i. e., $es_i = ee_i = $ null. Otherwise,

$$ee_i = es_i + d_i^{r_i}, \quad r_i \neq -1 \tag{3.10}$$

It is worth stating that if $es_i \neq$ null and $ee_i \neq$ null, then there must exist some resource j such that the execution in question starts at time es_i, end time ee_i, and the time window boundaries of the task ws_i^j, we_i^j satisfy a relation shaped as equation (3.11):

$$ws_i^j \leq es_i < ee_i \leq we_i^j, \quad \exists RS_j \in \mathbf{RS} \tag{3.11}$$

3.3.3 Objective function

The objective function is a measure of the quality of the scheme. Based on the assumptions, the objective function is a deterministic function of the decision variables, i. e., when a set of decision variables is fixed, the value of the objective function is uniquely determined. This research work does not consider modeling the imaging satellite task planning problem as a multi-objective optimization model for the following three main reasons:

① From the point of view of real needs, in actual engineering projects, decision makers usually focus on the most important objectives only, and although the Pareto front obtained by the multiobjective optimization method can reduce the decision space, for the imaging satellite controller and decision makers, what they ultimately want to get is a deterministic solution, not the solution set consisting of all solutions on the Pareto front [89, 90, 91, 65].

② From the point of view of multiobjective algorithms, regarding the solution of multiobjective problems, there are two mainstream solving ideas: linear weighting of multiple objective functions into a single objective function and hierarchical solution according to the importance of the objective function. These two ideas are ultimately multiobjective problems into single-objective solutions, so the actual problem is directly organized into a single-objective optimization model through the relevant technology, which can reduce the difficulty of model analysis and solution.

③ From the current state of research in the field of multiobjective planning, most of the current research in the field of multiobjective optimization is still at the theoretical level, and its application often needs to be business-driven, combined with specific application contexts to carry out the work.

Therefore, the authors believe that modeling the imaging satellite task planning problem as a multiobjective optimization model is both difficult and unnecessary under the current real-world conditions, so this work considers the problem as a single-objective optimization problem, with the objective function denoted by the symbol F. Typical objective functions and their basic forms in the field of imaging satellite task planning are summarized below:

(1) Maximize total profit from completing the task
Each imaging task has a certain value of profit after its completion. Maximizing the total profit of all completed tasks is the most intuitive and frequently applied objective function for the imaging satellite scheduling problem. Its basic form is as in equation (3.12):

$$\max F = \sum_{\{i|r_i>0\}} p_i^{r_i} \tag{3.12}$$

where $p_i^{r_i}$ denotes the profit value of task i executed in resource r_i.

(2) Maximize the number of tasks completed
When the profit of imaging tasks is difficult to describe quantitatively, the importance of different imaging tasks is difficult to compare between them. In this case, maximizing the number of completed tasks is used to measure the goodness of the scheduling scheme. When using this metric, each task is considered as equally importance, and thus, the objective function can be described as an expression shaped like the equation (3.13):

$$\max F = \sum_{\{i|r_i>0\}} 1 \tag{3.13}$$

(3) Maximizing robustness metrics for task planning schemes [92, 13]
In imaging satellite task planning scenarios, the robustness of the scheme is also a key concern for satellite controllers and decision makers in many cases. Because in many real-life imaging satellite task planning problems, a task planning scheme is maintained at all times, and when specific conditions are reached that require replanning in conjunction with a new situation, the decision maker often wants the new planning scheme to be as small as possible compared to the previous scheme:

$$\max F = \sum_{\{i|r_i>0\}} p_i^{r_i} g(es_i) \tag{3.14}$$

In equation (3.14), the objective function is obtained by summing the product of the profit value of each task planning and a penalization function $g(es_i)$, where the penalization function $g(es_i)$ measures the ability of the task i to resist perturbations. Depending on different practical considerations, this penalty function can have different forms, and is generally a function of es_i. Its range of values is usually $(0, 1]$.

(4) Maximize resource utilization
Resource utilization is the ratio that expresses the total profit to the resources consumed. The TotalConsumption in equation (3.15) represents the satellite resources to be consumed for the task planning scheme, which can be expressed in terms of the total consumed power, or other depreciation metrics can be designed to measure it:

$$\max F = \sum_{\{i|r_i>0\}} p_i^{r_i} / \text{TotalConsumption} \tag{3.15}$$

3.3.4 Constraints

The constraints of the imaging satellite task planning model also depend on the design of the specific satellite platform, application requirements, and other factors. After extensive research on the open literature and theoretical findings of the imaging satellite task planning problem [93], this study summarizes some general constraints for this problem as the inequality (3.16) to the inequality (3.23).

(1) Uniqueness
Each task can be executed at most once by a particular resource. This constraint is common for the vast majority of current imaging satellite task planning problems. According to the design of the decision variables in Subsection 3.3.2, for all nonempty variables r_i and es_i, the values of the variables at any moment are unique real numbers. Therefore, the design based on the output parameters of the model guarantees the uniqueness constraint of the scheme.

(2) Resource capacity
Equality (3.16) denotes that for any resource, the sum of storage space consumed by its scheduled tasks $\sum \text{Storage}(i,j)$ does not exceed the maximum available storage space of the current resource MaxStorage(j); equation (3.17) denotes that for an arbitrary resource, the total power consumed by the actions corresponding to its scheduled tasks $\sum \text{Energy}(i,j)$ does not exceed the maximum power available for the current resource MaxEnergy(j);

$$\forall j \in \textbf{RS}, \quad \sum_{\{i|r_i=j\}} \text{Storage}(i,j) \le \text{MaxStorage}(j) \tag{3.16}$$

$$\forall j \in \textbf{RS}, \quad \sum_{\{i|r_i=j\}} \text{Energy}(i,j) \le \text{MaxEnergy}(j) \tag{3.17}$$

(3) Imaging quality
Equality (3.18) indicates that all the tasks being planned need to take into account the imaging at the corresponding moment point for imaging when making decisions about

the imaging start moment es_i whether the clarity Quality(es_i) satisfies the minimum imaging quality requirement for the task i MinQuality(i);

$$\forall i \in \{i \mid r_i > 0\}, \quad \text{MinQuality}(i) \le \text{Quality}(es_i) \tag{3.18}$$

(4) Task visibility

Equation (3.19) and equation (3.20) denote respectively the execution time windows of all the tasks planned $[es_i, es_i + d_i^{r_i}]$ should be within the visible time window of the task:

$$ws_i^{r_i} - es_i \le 0 \tag{3.19}$$
$$es_i + d_i^{r_i} - we_i^{r_i} \le 0 \tag{3.20}$$

(5) Task conflictability

Equation (3.21) and equation (3.22) guarantee no overlap in the execution time window of any two tasks i_0 and i_1 do not have overlapping execution time windows. Equation (3.23) ensures that the imaging time interval between any two tasks is greater than their minimum imaging interval requirement. The function Trans(i_0, i_1) is the function to compute the imaging time intervals of tasks i_0 and i_1:

$$(es_{i_1} - es_{i_0})(es_{i_0} - ee_{i_1}) < 0 \quad \forall i_0 \neq i_1, r_{i_0} = r_{i_1} \tag{3.21}$$
$$(ee_{i_0} - es_{i_1})(ee_{i_1} - ee_{i_0}) < 0 \quad \forall i_0 \neq i_1, r_{i_0} = r_{i_1} \tag{3.22}$$
$$\text{Trans}(i_0, i_1) \le |es_{i_0} - ee_{i_1}| \quad \forall i_0 \neq i_1, r_{i_0} = r_{i_1} \tag{3.23}$$

In the above formulas, specific expressions for Storage and Energy in the resource capacity constraints, Quality in the imaging quality constraints, and Trans in the task conflictability constraints are usually related to the specific engineering design of the imaging resource, while the MaxStorage, MaxEnergy, and MinQuality can be considered as the expressions of the imaging resource whose value is determined before the scheduling process starts. The range of values for tasks and resources in all formulas are shown in equation (3.24) and equation (3.25):

$$i = 1, 2, \ldots, n \tag{3.24}$$
$$j = 1, 2, \ldots, mk \tag{3.25}$$

The above constraints are widely used in academic research on imaging satellite task planning problems, and constraints that are closely related to external conditions and practical applications (e. g., constraints on illumination conditions during imaging, constraints on payload temperature [94], constraints on the satellite's power consumption [95] and constraints related to satellite hardware capabilities [96]) can generally be lumped into one of the classes of constraints collated in this book to reduce uncertainty and facilitate task planning algorithms to deal with [96].

Starting from the typical engineering projects in the field of imaging satellite task planning, this model summarizes the forms of constraints for imaging satellite task planning problems into four major classes:

① **Cumulative constraints**. The use of specific physical statistics as constraints in the task planning model is very common in the practical engineering of imaging satellite task planning, which is of great practical significance to ensure the normal operation of resources and reduce the failure rate. The basic form of this class of constraints is "in a planning period, a certain statistic does not exceed a certain value," for example, "the cumulative duration of a satellite's daily imaging time does not exceed XX seconds."

② **Rolling constraints**. Rolling constraints control the peaks of certain statistics over successive periods of time. These types of constraints are generally used to ensure that payloads on a resource do not exceed the design criteria for a short period of time, thus serving to protect the payload [97, 96, 94]. This class of constraints is usually based on objective conditions, for example, in order to prevent damage to components caused by the high temperature of the satellite hardware, constraints such as "no more than XX seconds of imaging satellite in any consecutive XX seconds" are designed.

③ **Task attribute constraints**. This type of constraint limits the range of values of the decision variables of the task, for example, "single imaging time should not exceed XX seconds," "the imaging task can only be executed in the area with sufficient light conditions," and so on. Moreover, for each satellite and each type of task, the constraints on task attributes are diverse and have different forms of description: sometimes, they are standardized by mathematical expressions, and sometimes they are described by logical statements.

④ **Task correlation constraints**. Task correlation constraints ensure that there is no conflict between tasks in the final task planning scheme, i. e., it must be ensured that all the tasks being planned can be executed according to the plan under the given hardware conditions. This class of constraints mainly includes the time interval between two consecutive tasks, the logical sequential relationship between the imaging tasks and the data transmission tasks, and the logical relationship between different tasks.

By sorting out the assumption conditions and basic elements of the imaging satellite task planning problem, it is believed that the problem can be considered as a typical combinatorial optimization problem though, and the problem can be established as a corresponding mathematical planning model for processing. However, considering the special characteristics of the application background of this problem, its solution difficulty is much more incredible than that of the classical combinatorial optimization problem. The difficulty is mainly reflected in the following points:

① The hardware conditions of each imaging satellite are different, and their application models are even more varied. This leads to the poor generalizability of the satellite task planning models established in the literature. Further, the imaging satellite task planning problem, if the problem is described by a traditional mathematical planning model, the difficulties in subsequent simplification, deformation, and other treatments are extremely high.

② Using the traditional mathematical planning model to characterize the imaging satellite task planning problem, under the conditions of the explosive growth of users' requirements, the solution space will appears "combinatorial explosion." In this condition, using the rule-based reasoning operations research method or iterative local search-based metaheuristic algorithms are difficult to adapt to the development trend of rapid development of satellite response, thus making it difficult to achieve the solution process "fast, accurate, and stable," which is contrary to the goal of balancing the contradiction on solution quality and solution efficiency.

③ The refinement and diversification of user requirements necessitate that the solution to the imaging satellite task planning problem can be adapted to more complex application scenarios with strong generalizability. Therefore, it is increasingly difficult to manually design heuristic algorithms that can perform well in different complex scenarios based on the problem characteristics.

The goal of building the mathematical model in this study is that the model can be decoupled from the specific application context as much as possible to enhance its universality. Therefore, how to refine the essence of the satellite task planning problem and summarize the commonalities of the satellite task planning process under different working environments and constraints are the next focus of this chapter.

3.4 Problem decomposition and bilevel combinatorial optimization framework

Based on the boundary analysis of the problem, two solution processes of the imaging satellite task planning problem (the task assignment process and the task scheduling process) are defined, and a bilevel optimization model oriented to the imaging satellite task planning problem is established on this basis. According to the characteristics of the model, the solution framework of the problem is determined. Under this framework, the basic forms of the mathematical planning model for the task scheduling process and the MDP model for the task assignment process are established by combining the characteristics of the task planning model and its solution framework.

3.4.1 Problem decomposition

After an in-depth analysis of the imaging satellite task planning problem, combined with the description in reference [71], it was found that the problem can be described as two solution processes:

- On which resource should each task be executed?
- When does each task execute separately?

The two solution processes described above are distinctive, relatively independent, and closely interconnected:

(1) Once these two questions are answered, the corresponding imaging satellite task planning problem is solved.

(2) Solving the decision problems of the two processes sequentially in the order of task assignment followed by scheduling can effectively reduce the search space and simplify the solution process.

(3) Based on the above problem description, the context, constraints, and objectives of the imaging satellite task planning problem can be viewed as a black box that can be stripped from the mainstream process of decision-making. When the problem boundaries are defined, the mainstream process of decision-making may not change with these factors.

Combined with reflections on the problem of imaging satellite task planning, definitions of the two solution processes considered in this study are given, along with associated descriptions, which form the basis of this research work.

3.4.2 Imaging satellite task scheduling problem definition

Definition 3.2 (Task scheduling problem). Given n tasks on a particular resource j_0, the attributes of the task's imaging duration $d_i^{j_0}$, and the task's profit $p_i^{j_0}$ are known. Each task has a specific visible time window $[ws_i^{j_0}, we_i^{j_0}]$. Tasks are executed at different moments, and the impact on resource consumption and related tasks will change accordingly, and the specific impact mechanism is reflected through the constraints. Under the above conditions, the problem to be solved is: based on the determined task assignment scheme $\mathbf{r}$, consider all the constraints in the satellite task planning problem comprehensively, and decide the specific moment of execution for each task $\mathbf{es}$ and pursue the optimal of the objective function value for the satellite task planning problem $F(\mathbf{r}, \mathbf{es})$.

Regarding to Definition 3.2, the following points need to be clarified:

① The schematic diagram of the task scheduling problem is shown in Figure 3.4. In Figure 3.4, different types of rectangular boxes represent the visible time window of the task, the execution time window of the task, and the task that fails to be scheduled. The variables in the rectangular boxes represent the imaging duration of the task.

② The input of this problem is a combination of the input to the task planning problem and the output of the task assignment process, and the output of this problem is the time of execution of all scheduled tasks on each resource.

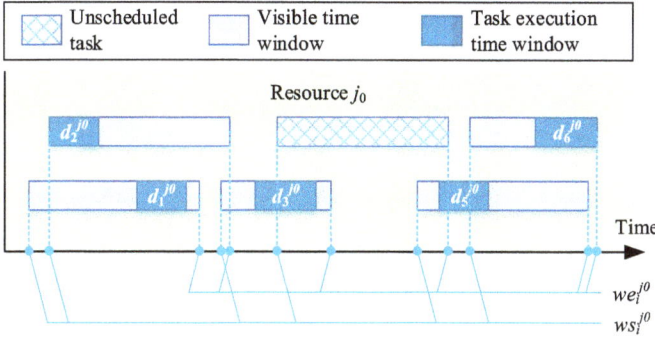

Figure 3.4: Schematic diagram of task scheduling problems.

③ The attributes of each task contain information about the visible time windows. The task scheduling process for imaging satellites that determines task execution time needs to consider the execution order of tasks first. The execution order of tasks with nonoverlapping visible time windows is fixed, i. e., if the end time of the visible time window of one task i_0 is earlier than the start time of the other task i_1, and both tasks are scheduled, then task i_0 must be executed before task i_1.

④ The task scheduling schemes on all resources comprise the final solution scheme for the imaging satellite task planning problem.

3.4.3 Definition of the imaging satellite task assignment problem

Definition 3.3 (Task assignment problems). Given n tasks, several imaging satellites, where there are k imageable orbiting cycles on each satellite. Each task may have multiple execution opportunities at multiple orbiting cycles, and each orbiting cycle may accept several tasks. However, each task is executed on at most one resource (a resource indicates a specified number of orbiting cycles in a specified satellite). Based on the above conditions and the constraints in the specific problem context, a feasible task assignment scheme is sought. The assignment scheme for the set of input tasks is represented by the vector $\boldsymbol{r} = (r_1, r_2, \ldots, r_n)$ and the task assignment process does not set a separate objective function. Its scheme is optimized according to the result of task planning.

Regarding to Definition 3.3, the following points need to be clarified:

① The task assignment problem is illustrated in Figure 3.5. In Figure 3.5, different colored boxes in the task set represent different tasks, and the number of resources is mk in total. The different types of rectangular boxes on each resource (i. e., each circle) represent the execution opportunities of the corresponding task. The number of rectangles is the number of execution opportunities.

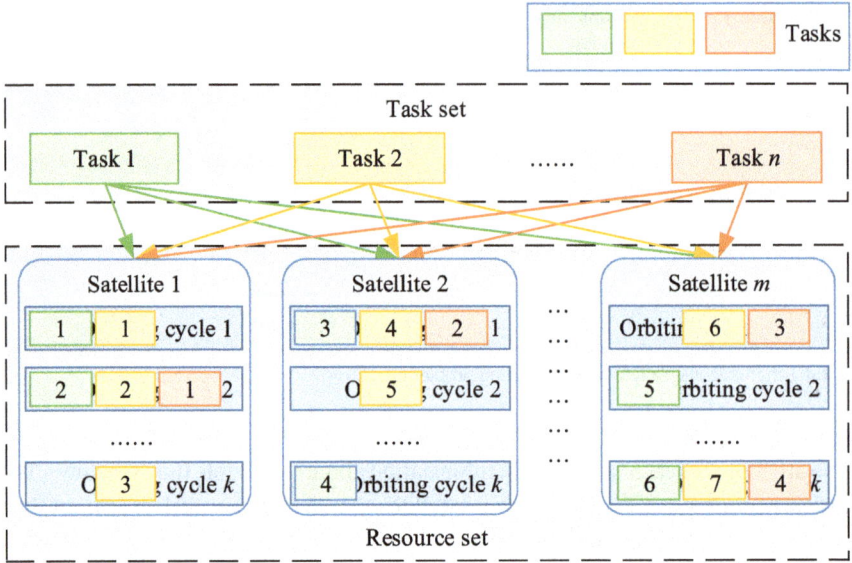

Figure 3.5: Schematic diagram of task assigning problems.

② The input to the model is information about the tasks and resources considered in the imaging satellite task planning problem, and the output is a scheme for matching each task and resource in the set of tasks.

③ A task does not necessarily have execution opportunities on all resources. Due to the combined constraints such as the geographic location of the task and the satellite's track, the set of optional resources for each task can be obtained through task pretreatment. Therefore, a task can only be matched with resources for which execution opportunities exist, which is the primary constraint to be considered in satellite task assignment problems.

④ Each resource has a certain capacity limitation and task number limitation (i. e., the total number of tasks that can be executed on each resource is limited), so if the task assignment is unreasonable, it will lead to a large number of tasks that can be accomplished by other resources not being accomplished, thus affecting the overall work efficiency of the system. It is possible to filter some of the obviously unreasonable assignment schemes by organizing the constraints of the task assignment process.

3.4.4 The framework of bilevel combinatorial optimization

By referring to the definition of the bilevel (also named "two-layer" in literatures) optimization model [41, 98, 99], the task assignment process, and the task scheduling process correspond to the solution of exactly two classes of decision variables, and thus can be naturally divided into two solution processes: the upper-level task assignment process realizes the decision on r, and the lower-level task scheduling process realizes the decision on **es**. In this section, a bilevel optimization model for the imaging satellite task planning problem is developed in the form of Model (3.26):

$$\min_{r,es} \quad F(r, es)$$

$$\text{s.t.} \quad G^u_{k_1}(r) \leq 0$$

$$G^l_{k_2}(r, es) \leq 0$$

$$k_1 = 1, 2, \ldots, g^u \quad\quad (3.26)$$

$$k_2 = 1, 2, \ldots, g^l$$

$$r = (r_1, r_2, \ldots, r_n) \in \mathcal{Q}_u$$

$$es = (es_1, es_2, \ldots, es_n) \in \mathcal{Q}_l$$

In the model, $F(r, es)$ represents the objective function; $G^u_{k_1}(r)$ and $G^l_{k_2}(r, es)$ represent the two constraint sets; $G^u_{k_1}(r)$ represents the constraint set considered in the upper-level task assignment process, $G^l_{k_2}(r, es)$ represents the constraint set considered in the lower-level task scheduling process; g_u and g_l represent the number of constraint entries in the upper-level and lower-level solution process respectively; r and **es** represent the decision variables; $\mathcal{Q}_u$ and $\mathcal{Q}_l$ denote the feasible domains, i. e., the solution spaces, of the corresponding decision variables in the upper-level and lower-level solution processes, respectively. The solution space of the imaging satellite task planning problem $\mathcal{Q}$ is the vector product of $\mathcal{Q}_u$ and $\mathcal{Q}_l$, i. e., $\mathcal{Q} = \mathcal{Q}_u \times \mathcal{Q}_l$.

3.5 Learning-based bilevel task planning model and thinking of solving

Based on the Model (3.26), this section analyzes the characteristics of the upper-level and lower-level solution processes, establishes the mathematical planning model and the MDP model, respectively, and analyzes the characteristics of these models, condenses the scientific problems that need to be focused on in the two models, and finally proposes a learning integrated solution framework, which is used to generally guide the design process of specific solution methods in the subsequent chapters.

3.5.1 Mathematical planning model for task scheduling

(1) Problem analysis

By referring to the design in Section 3.3, the inputs and outputs of the imaging satellite scheduling problem are shown in Figure 3.6. The input to the problem in Figure 3.6 is a

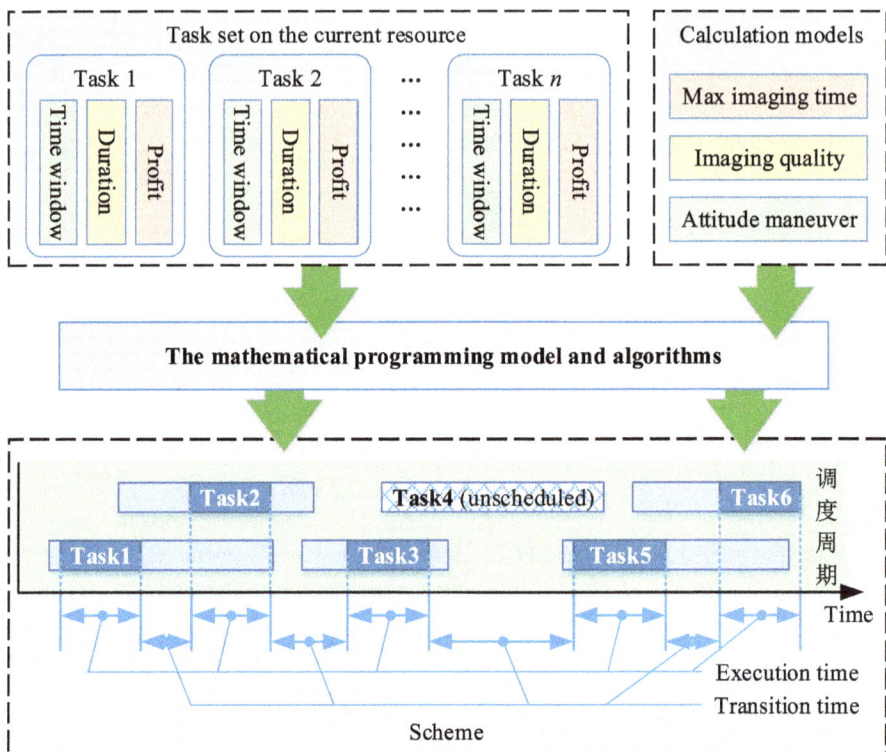

Figure 3.6: Schematic diagram of the input and output of the imaging satellite task scheduling problem.

collection of tasks received by the current resource, and each task in the collection has its fixed properties. The mathematical planning model of the task scheduling problem, which considering specific task collection and resource model, results in a complete task scheduling scheme that satisfies all the constraints.

This modeling approach can effectively reduce the complexity of the satellite task planning problem (i. e., Model (3.26)). In the Model (3.26), when the decision variables r are fixed for the upper-level task assignment process, the solution space for the lower-level task scheduling process is reduced. In the task scheduling problem studied in this book, it is assumed that the assignment scheme of each task has been determined, and each task is considered to be scheduled on only one of the corresponding resources. In other words, only one available time window is considered for each task in the task scheduling problem:

$$\Omega_l(r') = \{es \mid r = r', G^l_{k2}(r', es) \le 0\} \tag{3.27}$$

where r' is a set of values for the decision variables r.

The basic form of the task scheduling process can be schematized by the Model (3.28):

$$\min_{\textbf{es} \in \Omega_2(r')} \quad F(\textbf{r}', \textbf{es})$$
$$\text{s.\,t.} \qquad G^l_{k2}(\textbf{r}', \textbf{es}) \le \textbf{0}$$
$$k_2 = 1, 2, \ldots, g^l \tag{3.28}$$
$$\textbf{es} = (\text{es}_1, \text{es}_2, \ldots, \text{es}_n) \in \Omega_l$$

Therefore, the lower-level task scheduling process can be described as a combinatorial optimization model with lower complexity and targeted algorithms can be designed to solve it.

(2) Modeling

Based on the above analysis, this chapter using the agile satellite task planning problem (AEOSSP) as an example to help readers understand the modeling process of the task scheduling problems in the bilevel architecture. The AEOSSP built based on the literature [14, 100, 101] is more universally representative, and its constraints cover the basic constraints in the imaging satellite scheduling problem, such as the constraints on the uniqueness of the task execution, and the constraints on the visibility of the task. In addition to the above common general constraints, the model also has imaging clarity constraints and attitude transition time constraints, which are commonly mentioned in the imaging satellite scheduling problems, and can represent the individualized constraints of different models and types of imaging satellites. Through the analysis of these two kinds of constraints, it can be expanded to the simplification and processing of a class of constraints, which provides research ideas and methodological support for researchers in the field, and facilitates researchers to carry out standardized scientific research. The model is specifically described as follows:

$$\max_{\textbf{es} \in \Omega_2(r')} \quad F = \sum p_i, \tag{3.29}$$

s.\,t.

$$\text{ws}_i \le \text{es}_i \tag{3.30}$$
$$\text{es}_i + d_i \le \text{we}_i \tag{3.31}$$
$$\text{MinQuality}(i) \le \text{Quality}(\text{es}_i) \tag{3.32}$$
$$\text{Trans}(i, i+1) \le \text{es}_{i+1} - \text{ee}_i \tag{3.33}$$
$$\sum d_i \le \text{MaxDuration} \tag{3.34}$$
$$\sum \text{Storage}(i) \le \text{MaxStorage} \tag{3.35}$$
$$i \in \textbf{TS} \tag{3.36}$$

In this model, the tasks under study are all on the same resource, so the task and resource matching variable r_i in the model is simplified. The objective function is maximizing the total profit, as in equation (3.29). There are six main constraints considered in this model, covering the four classes of constraints designed in Subsection 3.3.4 of this book:

① Inequality (3.30) and inequality (3.31) denote the visibility constraints of the task, i. e., the start and end times of the task execution need to be within the range of the visibility window. Inequality (3.32) denotes the imaging quality constraint, i. e., the imaging quality corresponding to the moment of task execution has to meet the imaging quality requirements. The task attribute constraints represented by inequality (3.30) to inequality (3.32).

② Inequality (3.33) denotes the transition time constraints between a task and its neighboring tasks, which belongs to the task correlation constraint.

③ Inequality (3.34) expresses the maximum imaging duration constraint, i. e., the total imaging duration of the imaging tasks during the current orbiting cycle (one orbiting cycle of about 90 minutes for most LEO satellites) cannot exceed the maximum allowed imaging duration. This constraint is a type of rolling constraint.

④ Inequality (3.35) denotes the storage constraint, i. e., all the imaged tasks need to occupy a solid-state memory that does not exceed the maximum allowable solid-state memory of the current resource. This constraint belongs to the cumulative constraint.

The model is reasonably simplified based on actual engineering problems, which facilitates theoretical research and retains the essential features of engineering problems. This book will summarize a class of constraints by processing and analyzing some of the constraints in Chapter 4 and realize the efficient solution of this model by designing constraint checking algorithms and deterministic algorithms in a targeted way.

(3) Scientific issues

Solving this problem needs to ensure that all the constraints are satisfied. As the constraints become more complex, using relatively general and efficient algorithms to ensure the quality of the solution is at the heart of the study of the task scheduling problem in this book.

The scientific issues to be addressed in the imaging satellite task scheduling problem are threefold:

① Unified modeling of constraints. Taking "SuperView-1" commercial remote sensing constellation as an example, there are around 70 constraints for each satellite in the constellation, and the task planning process for other satellites may need to consider more and more complex constraints. Therefore, how to unify the processing of constraints in different types is the first difficult problem to be solved in the task scheduling problem of imaging satellites.

② Design of efficient constraint checking algorithms. The constraint checking algorithm, as an underlying algorithmic tool, is usually nested in each scheduling algorithm. Therefore, the computational efficiency of constraint checking algorithm has a deep impact on the solution efficiency of the whole scheduling algorithm. How to ensure the completeness of constraint checking and improve the computational efficiency of constraint checking is another difficulty in solving the task scheduling problem.

③ Balancing the universality, computational efficiency, and solution quality of the solution algorithm in the imaging satellite scheduling problem. In the task scheduling problem, to improve the universality of the algorithm, it is necessary to decouple the algorithm from the constraints, and to take into account the computational efficiency and solution quality, it is necessary to skillfully design the algorithm operation rules according to the problem characteristics to reduce the ineffective computational process. It is the pursuit direction of the task scheduling problem solving, but also the biggest difficulty in the problem solving process.

The research ideas and objectives of the task scheduling problem are: to establish a mathematical planning model of the task scheduling problem, to propose an efficient constraint checking algorithm by classifying and analyzing the constraints, to design the task scheduling algorithm based on the constraint checking algorithm in order to realize the decoupling with the specific constraints, and to theoretically prove the optimality of the algorithm by calculating the complexity of the algorithm and through mathematical derivation, so as to ensure that the algorithm in imaging satellite scheduling problems with universal applicability, high efficiency, and high solution quality.

The research idea and solution process of the task scheduling problem, as well as the approach to the problem difficulties and experimental verification are developed in detail in Chapter 4 of this book.

3.5.2 MDP model for task assignment

(1) Problem analysis

Unlike the task scheduling problem, the task assignment problem is not suitable to be built into a mathematical planning model for processing. This is because the task assignment problem considered in this study does not have a separate subobjective function to evaluate the goodness of the task assignment scheme. Therefore, the result of the task assignment problem needs to be determined with a specific scenario to calculate the objective function of the task planning scheme after a series of operations in the task scheduling process $F(r, es)$. In the research of similar problems, many scholars try to manually set subobjective functions for the task assignment problem, thus transforming the task assignment problem into a relatively independent combinatorial optimization model. This approach can get better solution results in a simple problem, but as the problem becomes more and more complex, it is difficult to guarantee the reasonableness of the designed subobjective function by processing the model in this way, so this study adopts a new way of thinking about this problem.

Theoretically, if the task scheduling algorithm is a deterministic algorithm, there is a unique objective function value F corresponding to it for any task assignment scheme. However, due to the complexity of the solution process of the task scheduling algorithm, it is often impossible to find an explicit function to determine the relationship between the task assignment scheme and the overall objective function value of task planning. How to utilize the means of machine learning to train under the condition of limited computational resources to obtain an empirical formula for representing the relationship between the task assignment scheme and the objective function value of task planning is one of the key issues in the study of the task assignment problem.

The MDP model used to solve the imaging satellite task assignment problem has advantages unmatched by other mathematical models:

① The purpose of the MDP model is to obtain an empirical formula for guiding task assignment through iterative training in a constructed simulation environment. Once the operating rules of the simulation environment have been determined, the process of training can be carried out using reinforcement learning algorithms, a process that does not require the preparation of labeled data in advance or the manual design of decision criteria.

② The task assignment problem can be described as a sequential decision-making problem, i. e., tasks are added to the solution scheme one by one in a sequential order, and the next decision is made according to the change of the scheme. The MDP model is a class of mathematical models used to solve sequential decision-making problems, which can fit well with the essential characteristics of the problem.

③ It is difficult to consider scheduling-related constraints in a task assignment problem, but the specific form of the constraints can have a great impact on the final result. The MDP model realizes the decoupling of the model and the actual constraints by fitting the characteristics of the constraints through the comprehensive analysis of the profit values of the feedbacks after taking actions.

④ After the empirical formula has been trained, the time complexity of using the empirical formula to solve the task assignment problem is low. The design of the training process is used to ensure the quality of the task assignment so that the efficiency and solution quality of the task assignment process can be guaranteed.

In addition, there are many decision variables in the imaging satellite task planning problem, but the evaluation of these decisions is based on the final scheduling scheme, and the impact of each step of the decision on the whole system is intrinsically related to the good or bad of the final scheduling scheme. The MDP model is able to record the instantaneous profit of all the decisions and summarize the intrinsic connection between the instantaneous profit and the long-term value of each time step, so as to guide the decision-making. The above features illustrate the superiority of modeling the task assignment problem as an MDP model compared to other mathematical models. However, modeling a real-world problem as an MDP model is a skillful task. The same real-world problem can be modeled as different forms of MDPs, and the form of MDPs has a fundamental impact on the solution process, so finding an appropriate MDP model will make the solution process twice as effective.

The published MDP models for planning and scheduling can be categorized into two main classes: "end-to-end" and "step-by-step" models.

The "end-to-end" model means that a complete solution can be directly generated by the agent in the MDP model through a single action. This approach is an emerging solution idea in recent years, and has been successfully applied to the convex set problem [102], the VRP problem [39], etc. Its core idea is to input a collection of tasks to be decided upon by inputting them to a deep learning network, and then output the arrangement of this collection of tasks in one step. When specific deep neural networks such as pointer networks are combined with the reinforcement learning process, their features are just right for solving the hardness of the planning and scheduling problem, so they have attracted a lot of attention. The outline of the basic elements of the design of such MDP models is summarized in Table 3.1.

Table 3.1: Essential elements of the "end-to-end" MDP model.

Elements	Description
Action	Generate a sequence of tasks
States	All Task Properties
Short-term reward	Total profit under current task sequence
Long-term value	Projected total profits

The model is simple to describe, the training process is easy to understand, and its effectiveness depends largely on the structure and parameter design of the pointer networks. However, from the current open research, the MDP model built based on this strategy usually has slow convergence speed, low learning efficiency, and easy to fall into the local optimum during the solution process, so it is usually considered to be used in the application scenarios in which the computer hardware resources have reached a certain level or a longer training period is allowed.

The "step-by-step" MDP model is a more traditional thought, and its basic process is: starting from the initial state, the agent gets a single-step decision based on the current state. Based on the change in state and the short-term reward obtained after the single-step decision, the long-term value function is updated, and the subsequent decision is made. This cycle continues until the termination condition is reached. For ease of understanding, the process can be represented as the Figure 3.7.

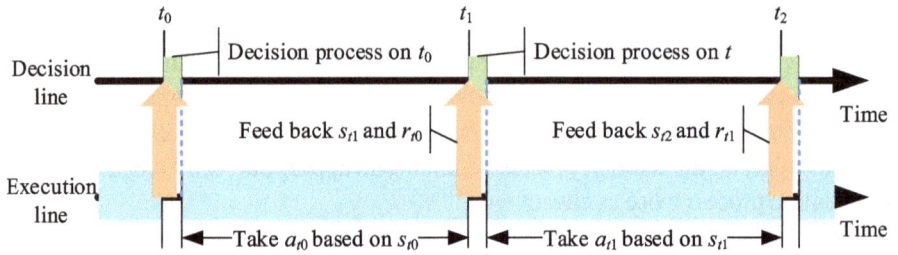

Figure 3.7: "Step-by-step" MDP model interaction.

Table 3.2: Essential elements of the "step-by-step" MDP model.

Elements	Description
Action	Accept or reject the current task
States	Satellite operating status, current task attributes
Short-term reward	The reward gained by current action
Long-term value	projected total profits

This model is more commonly used in planning and scheduling than "end-to-end" models. Its basic elements are summarized in Table 3.2.

(2) Modeling

Each of the above models has its own advantages and disadvantages: "end-to-end" models are very efficient to apply after training, because it can generate the solution directly with one-step actions. However, the training process usually consumes more computational resources and time, and it is easy to fall into local optimization; "step-by-step" models have a relatively smoother training process, and it is not easy to fall into local optimization. Therefore, this study tries to model the imaging satellite task planning problem using a "step-by-step" model, and the basic elements of the designed MDP model are shown in Table 3.3.

Table 3.3: Essential elements of the MDP model for the imaging satellite task assignment problem.

Elements	Description
Action	Task to be performed by the current resource
States	Properties of all tasks
Short-term reward	Difference on total profits between current and previous states
Long-term value	projected long-term returns

Combined with the analysis of the bilevel optimization model for imaging satellite task planning, the MDP model for the task assignment problem can be briefly described by Figure 3.8.

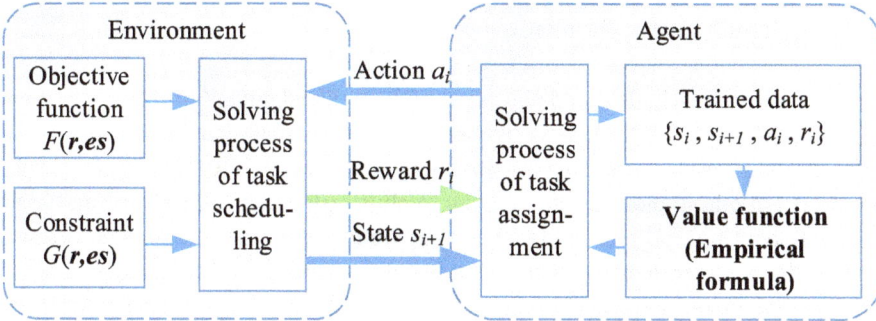

Figure 3.8: Conceptual model for solving the imaging satellite planning problem.

Based on the reflections on task decomposition in Subsection 3.4.1 of this chapter, the imaging satellite task planning problem can be solved in two solution phases. Initially, the system is in state s_0. The agent chooses action a_0 based on s_0. The environment reads the decision of the agent a_0 and combines it with the current state s_0 to compute the short-term reward r_1 of the action a_0; at the same time, the state of the next step s_1 is updated. In this cycle, the agent generates new actions a_i based on s_i, and the environment receives an action and feeds back the new state and reward, which are used as the basis for the training of the agent and the subsequent decision-making until the final state is reached.

In Figure 3.8, the solution process of the task scheduling problem is considered to be a part of the environment in this MDP model and is designed in accordance with the solution thinking in Subsection 3.5.1. The deterministic algorithm is implemented, while the solution of the task assignment problem is realized using a trained value function. This value function is obtained from the cyclic interaction between the environment and the agent, which is constantly updated. After many training iterations in different scenarios, the value function eventually converges and can play its guiding role in the decision-making process of the task assignment problem.

(3) Scientific issues

The three main scientific issues that need to be addressed in the imaging satellite task assignment problem are as follows:

① How to reasonably design the elements in the MDP model to narrow the search space and improve the training efficiency and training accuracy based on the consideration of the characteristics of the task planning problem for imaging satellites.

② How to take full advantage of the stochastic nature of imaging satellite task planning scenarios to design strategies that trade-off convergence and generalization of algorithms.
③ How to make full use of domain knowledge to design a reasonable action pruning strategy and realize the algorithm to select actions more efficiently and reasonably during training and application.

Therefore, the research ideas and objectives of the task assignment problem are: to establish the MDP model of the task assignment problem and to design the elements in Table 3.3 in detail based on the characteristics of the imaging satellite task planning problem; to design and implement the imaging satellite task assignment algorithm based on deep Q-learning, and realize the pruning operation of the task assignment process based on domain knowledge to improve the training efficiency and application effect of the algorithm; to verify the performance indexes such as convergence, generalization, solution efficiency, and solution accuracy of the algorithm through simulation experiments. The DQN algorithm is combined with two deterministic algorithms that have their own advantages and disadvantages to form an integrated algorithm for solving the imaging satellite task planning problem. The advantages of the designed algorithm are further demonstrated by comparing its performance with other reinforcement learning algorithms (asynchronous advantage actor-critic algorithm, pointer network-based actor-critic algorithm) in gradient experiments.

The research ideas about the task assignment problem, the solution process, the approach to the difficulty of the problem, and the experimental validation are developed in detail in Chapter 5 of this book.

3.5.3 Learning-based task planning technique

This study proposes a learning-based bilevel task planning technique for imaging satellites, whose solution framework is shown in Figure 3.9. In this framework, the imaging satellite task planning problem is divided into two levels of optimization processes and appropriate algorithms are designed to solve them, respectively. The upper-level task assignment problem is solved based on the empirical formulation obtained from the training of the reinforcement learning algorithm. During the training process, the task assignment problem is modeled as an MDP model, and sufficient and diverse training data are obtained by assigning tasks in different scenarios for fitting the required empirical formulas; the task scheduling problem is modeled as a mathematical planning model in order to design deterministic algorithms to construct a satisfactory imaging satellite scheduling scheme under limited computational resources. The design of deterministic algorithms is necessary for the training process of the MDP model. The solution processes of the two problems need to interact continuously during the training of the task assignment agent. In the task assignment module, the algorithms take into

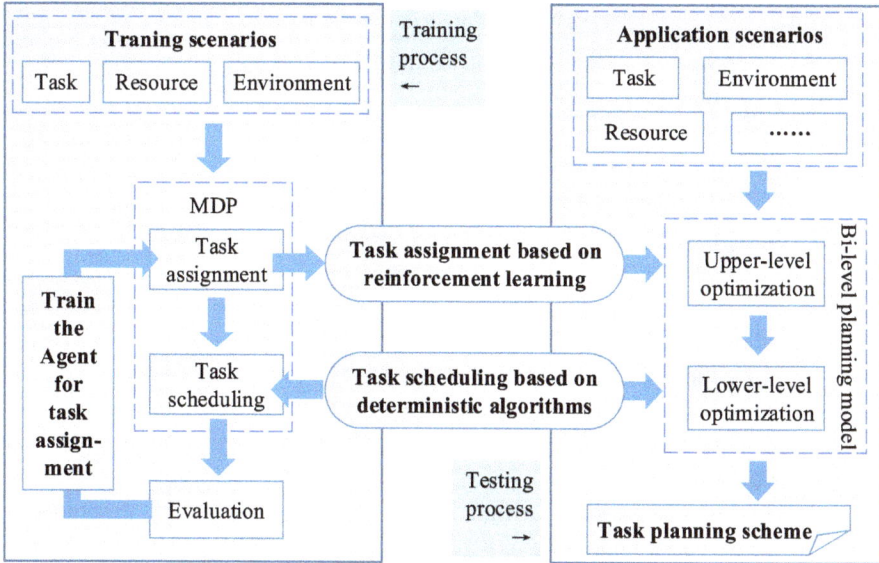

Figure 3.9: Technical framework for learning-based bilevel task planning.

account the decision-making preferences of the agent, and "Exploration" and "Exploitation" strategies are adopted to make decisions. The decision result is used as the input of the task scheduling process, which calculates the optimized objective value of the final scheduling scheme according to the received task assignment scheme, and uses the reinforcement learning algorithm to modify the decision preference of the agent to achieve the training of the empirical formula of task assignment.

After the model is trained, the empirical formula is used as the decision basis for solving the task assignment problem, and a reasonable task assignment strategy is selected according to the empirical formula when different tasks are input. Combined with the deterministic algorithm for solving the task scheduling problem, the final imaging satellite task planning scheme can be obtained quickly and accurately. This is a new paradigm for solving the task planning problem of imaging satellites that minimizes the coupling of specific constraints in the solution process and maximizes the generality of the algorithm. The scheme combines reinforcement learning algorithms with deterministic algorithms in operations research and gives full play to the advantages of both: the deterministic algorithms can obtain stable and satisfactory solutions in polynomial time, while the reinforcement learning algorithms can show strong generalization ability and high solution efficiency in unknown scenarios at the end of training. The method can also be transferred to solve other similar problems quickly, which has great application prospects and research value in facing increasingly complex practical problems in the future. Pseudocodes for the basic process of reinforcement learning training and testing are given in Algorithm 3.1 and Algorithm 3.2.

Algorithm 3.1 Training process in RL for task assignment.

Input: Global information such as task collections, resource collections, etc.

Output: Value function

  1: **repeat**

  2:    Read a new task planning scenario;

  3:    **while** Does not met terminal conditions **do**

  4:       Parameter initialization;

  5:       **while** Does not reach the terminal state **do**

  6:          Selects an imaging task for the current resource based on the characteristics of scenarios.

  7:          Adds the task to the set of tasks to be scheduled for the current resource and uses a deterministic algorithm to obtain a task planning scheme.

  8:          Update the status after adding the task based on the new task planning scheme.

  9:          Trains a long-term value function based on the selected task, the states before and after updating, and the short-term reward of the action;

10:       **end while**

11:       Assigns tasks for the next resource and resets the resource state;

12:    **end while**

13: **until** Trained in all scenarios

3.6 Summary

Based on the analysis of the satellite operation and control process and the design of the satellite task planning system, this chapter focuses on the modeling of the imaging satellite task planning problem and its related work. First, it defines the imaging satellite task planning problem, puts forward the basic assumptions of the problem study, overviews the modeling ideas of researchers in the field of imaging satellite task planning for the problem over the past decades, and summarizes the advantages and disadvantages of various types of models as well as the development trend.

    Then the basic elements of the imaging satellite task planning problem (input, output, objective function, and constraints) are discussed in detail, and the boundaries of the problem are standardized. Based on the comprehensive analysis of various types of constraints and problem characteristics, the problem is decomposed into task assignment process and task scheduling process, and a bilevel optimization model for imaging satellite task planning problem is established. The model describes the decision-making in the task assignment process as the selection of the visible time window of the task, does not distinguish between the single imaging satellite task planning problem and the multisatellite collaborative task planning problem, and can uniformly describe the task planning problems of several common classes of imaging satellites as well as the

---

Algorithm 3.2 Testing process in RL for task assignment.

Input: Global information such as task set, resource set, trained value function
Output: Task planning scheme
  1: Input application scenarios;
  2: **while** Does not reach the terminal state of current scenario **do**
  3:    Parameter initialization;
  4:    **while** Does not reach the terminal state of the current resource **do**
  5:        Select the task with the highest value to add to the current resource;
  6:        Add the task to the set of tasks to be scheduled for the current resource.
  7:        Compute for the updated task planning scheme by the task scheduling algorithm.
  8:    **end while**
  9:    Assigns tasks for the next resource and resets the resource state;
10: **end while**

---

heterogeneous multisatellite task planning problem, to realize the standardization and unification of the model description.

Finally, based on the unified bilevel optimization model, the research framework of the learning-based bilevel task planning technique is proposed, which is the methodological basis for the subsequent research work. The deterministic algorithm is designed to solve the task scheduling problem, and the reinforcement learning method is designed to solve the task assignment problem, which is realized by the idea of "exploration-scheduling-learning" in the solution framework. Following this idea, the two parts of the decision-making process in the solution framework are loosely coupled, so as to reduce the complexity of the task scheduling process and improve the rationality of the task assignment process.

4 Research on task scheduling problems of imaging satellites based on deterministic algorithms

This chapter focuses on the nature of the imaging satellite task scheduling process and the solution method, the difficulties that lie in how the algorithm handles various complex constraints, and the goal of ensuring the algorithm's solution efficiency and solution accuracy under complex constraints. First, the two common classes of complex constraints in the imaging satellite scheduling problem are analyzed to give the idea of eliminating the corresponding types of constraints in the actual engineering problem. Then the constraint checking method based on the timeline advancement are proposed. Based on this, two deterministic algorithms for scheduling imaging satellites are proposed: heuristic algorithm based on the density of residual tasks (HADRT) and dynamic programming algorithm based on task sequencing (DPTS). The implementation process and optimality proofs of the two methods are given in the paper, illustrating the solution efficiency and effectiveness of the algorithms. Through the theoretical analysis and experimental comparison of these two algorithms, the performance differences between these two algorithms in specific scenarios are explored, so as to provide suggestions for algorithm selection for different scenarios.

4.1 Literature review

There are three classes of mainstream imaging satellite task scheduling algorithms: exact algorithms, heuristic algorithms, and metaheuristic algorithms. There are fewer studies related to the direct application of machine learning to imaging satellite task planning, so it will combined with the introduction to the application of machine learning in the field of combinatorial optimization in Subsection 5.1.2 of this book.

In the imaging satellite task scheduling problem, widely used exact solution algorithms include branch and bound [103], branch and pricing [104], and dynamic programming [105] algorithms. In addition, Bistra Dilkina [106] use a combination of depth-first search and constraint propagation to achieve the solution of the agile satellite planning and scheduling problem. These classes of algorithms can guarantee an optimal solution in finite time in small-scale task scheduling scenarios or simplified problems, but the computational time is often unacceptable when solving large-scale problems or complex problem contexts, due to the complexity of the problem model itself.

Construction heuristic algorithms are commonly used algorithms in this field. Song Liu et al. [86] proposed a rolling rule heuristic algorithm combining random mechanisms and roulette ideas for the autonomous task planning problem of agile imaging satellites, where the search process can be understood as an iterative process, and each iteration consists of two phases: a construction phase and a local search phase, which accompany the process of each iteration. However, this class of algorithms is not very

https://doi.org/10.1515/9783111585109-004

popular in general practical problems, the main reason is that this kind of heuristic algorithm has randomness, and the stability of the solution is difficult to guarantee when the number of iterations cannot reach a certain scale. Xue Zhijia et al. [107] proposed a two-phase planning algorithm combining heuristic search and improved plan review technique, which can quickly and effectively solve the satellite autonomous task scheduling problem under unexpected conditions; Pierre F. Maldague [108] et al. introduced the concept of JIT in industrial engineering to design heuristic rules to realize satellite task scheduling, emphasized the relative time relationship between tasks, and continuously adjusted the timing of satellite observation tasks in real-time according to the advancement of the actual process, in order to continuously search for the optimal execution time and improve the observation efficiency. In small-scale and medium-scale imaging satellite scheduling scenarios, although heuristic algorithms are poor in terms of solution accuracy, their solution speed is fast, and they can obtain a feasible solution in a short computing time.

It is because of the low computational complexity, controllable computing time, and strong interpretability of the construction heuristics that heuristics are widely used in practical engineering, such as the EO-1 [109] in the United States, the Pleiades [110, 111, 112] in France, and the FireBIRD [113] in Germany, among others. Table 4.1 summarizes a selection of representative imaging satellite application projects in the United Station, European Union, France, and Germany, and their proposed scheduling algorithms for imaging satellites, from literatures.

Table 4.1: Typical engineering projects and their scheduling algorithms in the field of imaging satellite task planning.

Number	Pilot project	Affiliation	Scheduling algorithm
1	Automated Scheduling and Planning Environment (ASPEN) [109, 114, 115]	NASA, the USA	Iterative repair algorithms
2	Pleiades' Task Analysis and Planning System [116]	CNES, France	Stochastic greedy algorithm
3	The Verification of Autonomous Mission Planning Onboard a Spacecraft (VAMOS) [113, 117]	DLR, German	Residual resource based heuristic algorithms
4	The Project for On-Board Autonomy (PROBA) [118, 119]	ESA, the EU	Heuristic algorithms based on cost functions

(1) EO-1: Iterative repair algorithm

CASPER, designed by Steve Chien et al., accepts goal-based commands and schedules a series of actions to reach a goal state without violating any rules or constraints. CASPER is able to generate planning schemes using an iterative repair-based local search algorithm and continuously planning tasks on the schemes during execution [120]. First, the algorithm generates an original scheme that may have conflicts or constraint violations, and attempts to resolve one conflict at a time until all conflicts have been resolved. By applying the iterative repair algorithm, continuous planning can be achieved, i. e., instead of planning tasks periodically, an immediate response is achieved and the execution results are incorporated into the planning process, as shown in Figure 4.1. For the generated planning scheme, if anomalies or new science events are found, an iterative repair algorithm is also used for the search of feasible solutions. The search space of the algorithm is huge due to the existence of multiple conflict types and solutions onboard. EO-1 takes heuristic rules to improve the search efficiency of the algorithm, and the algorithm prefers to choose the rules with a high confidence level.

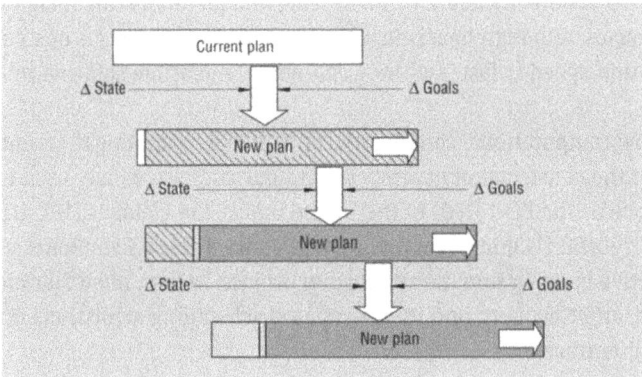

Figure 4.1: Iterative repair algorithm of "EO-1" [120].

(2) Pleiades: Stochastic greedy algorithm
Gregory Beaumet proposed an on-board stochastic greedy algorithm for Pleiades, with a stochastic decision model as shown in Figure 4.2. The process can be described by a decision tree model, where the rightmost node of each layer represents an action that can be realized by the satellite, and its order is ranked from top to bottom by a predefined priority, i. e., when multiple actions can be executed at a certain point in time, the action represented by the node of the top layer is considered first, and is checked sequentially downwards. In the graph, some nodes are random nodes, i. e., it is random to choose a branch under that node. For example, there are three different heuristic rules to schedule the imaging satellite's task, and the exact choice of which rules to use for planning is determined by the value of a random variable. The final action performed is influenced by the three parameters p, q, and r. According to this algorithm, a feasible satellite action can be obtained every time an action planning is needed.

(3) FireBIRD: A task planning algorithm based on real-time constraint checking [88]
In VAMOS, the purpose of global task planning on the ground is to utilize efficient computation as much as possible to maximize the profit and reliability of the planning scheme.

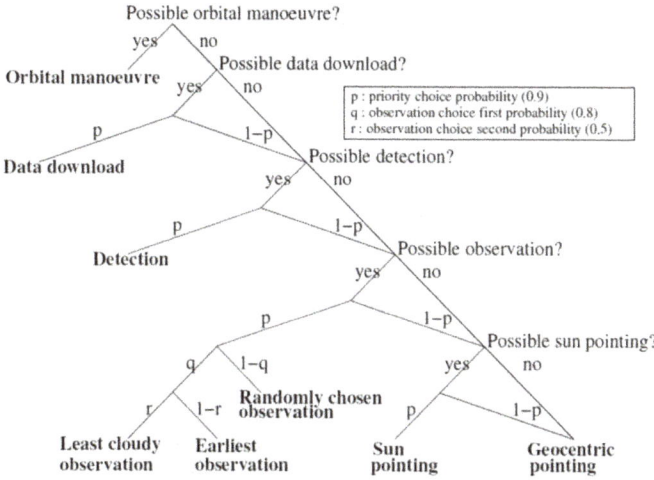

Figure 4.2: Stochastic greedy algorithm of "Pleiades" [88].

After each command is executed on the satellite, the on-board event triggers the timeline plug-in OBETTE to check and update the constraints and resources information, and if it is found that the gap between the predicted and actual values of the constraints and resources is not within the acceptable range, the planning will be carried out immediately. An invariant prioritized greedy rule is used for planning on-board with the aim of reducing the complexity of the software system on the satellite. Figure 4.3 represents the process of on-board time window selection, which is realized by comparing the actual constraints on-board and the prediction environment information on the ground. The on-board time window is selected and adjusted, and guides the subsequent task planning. For each on-board capability (power, solid-state memory, etc.), a similar graph is obtained, and repeated calculations can be avoided during constraint checking.

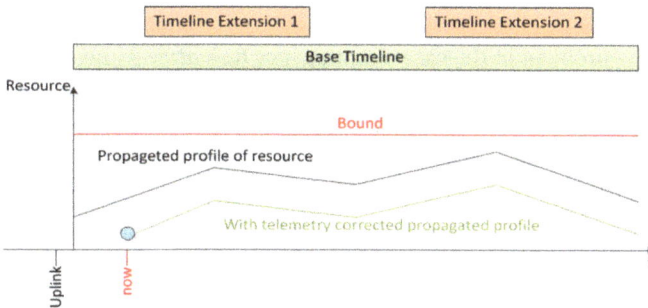

Figure 4.3: Real-time constraint checking-based task planning of "FireBIRD" [113, 117].

By analyzing the typical engineering applications in Table 4.1, combined with the current status of algorithm application in Chinese imaging satellite scheduling system, it is not difficult to find out that: heuristic algorithms have the advantages of clear algorithmic structure, reliability of the solution results, and low computational complexity, which are popular in practical engineering problems with high requirements of timeliness and reliability.

The idea of metaheuristics for solving decision problems originated in the 1950s [40], while it has been widely used in the field of imaging satellite task planning in the last decade. Adaptive large neighborhood search metaheuristic algorithms (ALNS) [14], genetic algorithms [77, 121], and ant colonies and their variants [122] have been considered in previous studies for scheduling imaging satellites. In addition to this, Globus et al. [123, 124] compared genetic algorithms, hill-climbing algorithms, etc., to design a problem model and solve it for different task sizes; D. Jamal Habet et al. [125] improved the tabu search algorithm by incorporating incomplete enumeration into the algorithm increases the speed of solving agile satellite planning problems and ensures consistency of results through the propagation of constraints; in some cases, these algorithms obtain better results than some mathematical algorithms and heuristic algorithms, but two limitations make them impractical for real-world problems: (1) it is difficult to find generalized parameters and functions that can produce good results for different inputs and (2) in complex cases, it is difficult for these types of algorithms to converge in an acceptable time.

Overall, whichever of the three methods described above is used, their decision rules are human-made and do not change during the calculation process. This leads to the fact that these rules may perform poorly once the context of the problem changes.

For the imaging satellite task scheduling problem under specific conditions (such as the "dynamic task scheduling problems" or "task replanning problems"), many experts and scholars in China have modeled the problem from different perspectives and proposed suitable algorithms to solve it, and these works have inspired the related research work in this book. Liu Yang [84] describes the dynamic task scheduling problem of imaging satellite as a class of CSP model and solves it, but the satellites under consideration are non-agile satellites; Zhai Xuejun [126] proposes a robust planning model to describe the task planning problem in the case of emergency task arrival and designs a multiobjective genetic planning problem and design a multi-objective genetic algorithm to solve the problem, but this method cannot obtain a stable solution in a limited time, which does not meet the requirement of high reliability on-board the satellite; Wang Maocai [127] designed two heuristics to realize the task planning, which is able to achieve high time efficiency under the condition of large-scale inputs, but due to the fact that their heuristic rules are too simple, resulting in poor solution quality; Jianjiang Wang et al. [128] designed dynamic task merging strategies and a novel real-time task planning algorithms for real-time contingency task arrivals in a multisatellite system. Baocun Bai [105] considered the problem as a knapsack problem and solved it using a dynamic programming approach; Zhenzhen Yan [129], Feng Yao [130], Lining Xing et al.

[131, 132] used an intelligent optimization algorithm represented by the ant colony algorithm or an improved knowledge-based intelligent optimization method to solve the agile satellite imaging task scheduling problems, which makes full use of the data generated during the solution process to enhance the computational efficiency and improve the quality of the solution, and provides a new way of thinking for solving this type of problems.

4.2 Constraint analysis and constraint checking

A generalized and efficient constraint analysis and checking process can lay a solid foundation for the imaging satellite task planning solution process. Based on the understanding of the imaging satellite task planning problem and the analysis of the mathematical modeling process, the necessary concepts are first defined, and then the processing ideas of a class of constraints are given as an example of imaging quality constraints and attitude transition time constraints; then the constraint checking algorithm based on the timeline advancement mechanism is proposed to realize a unified and fast checking process for different types of constraints; in this constraint checking algorithm, the heuristic algorithm based on the density of residual tasks (HADRT) and the dynamic programming algorithm based on task sequencing (DPTS) are designed on the basis of this constraint checking algorithm, and the efficiency and optimality of these algorithms are theoretically proved. At the same time, the application effect of the algorithm in specific scenes is verified through simulation experiments.

4.2.1 Related concept definition

In Model (3.26), the matching scheme of resources and tasks is determined (i. e., r' being viewed as constants) under the condition that the task scheduling problem is characterized separately. Before analyzing the problem, a few definitions need to be introduced.

Definition 4.1 (Task set). In the task scheduling problem, the tasks in the task set are the subject of the decisions to be made. The set is finite, i. e., the number of tasks considered for each task scheduling is limited. Each object (task) in the task set is an independently existing individual, and each task has a number of attributes, including variables that have been identified in the input conditions and decision variables.

The set of tasks can be represented and processed by equation (3.6), equation (3.7), and equation (3.8). Task sets have the following properties: determinism of elements, mutual dissimilarity of elements, and disorder of elements. To simplify the representation, the task i is usually referred to in mathematical models by TS_i.

Definition 4.2 (Task sequence). The set of objects obtained by arranging the objects in the task set in a particular order is called task sequence. Since the task set is a finite set, the number of objects in the task sequence is also finite. A task sequence is uniquely determined once the logical sequential relationships of the objects are determined, and there are no juxtaposed objects in the task sequence. A task sequence is usually represented by an n-dimensional vector: $(tsq_1, tsq_2, \ldots, tsq_k)$, $tsq_i \in \textbf{TS}$.

Task sequences are usually used as intermediate products for the solution of task scheduling problems, and are of great value in the efficient solution of task scheduling problems. According to the analysis of the operation process of imaging satellite in this study, the qualifying tasks have unique constraints, so the tasks in the sequence are not repeated. In the imaging satellite scheduling problem, the complete task sequence is a special kind of task sequence whose length is equal to the number of elements of the task set, denoted as **TSQ**, as shown in equation (4.1) and equation (4.2):

$$\textbf{TSQ} = (tsq_1, tsq_2, \ldots, tsq_n) \tag{4.1}$$

$$length(\textbf{TSQ}) = card(\textbf{TSQ}) = n \tag{4.2}$$

where function length(**TSQ**) is used to find the length of the task sequence **TSQ**, and the function card is used to find the base number (i. e., the number of elements of the set) of the set **TSQ**.

The realistic meaning of a complete task sequence in the imaging satellite task scheduling problem is a sequence based on some algorithm or rule that takes into account all the objects in the task set. The processing of the complete task sequence during the subsequent operations only requires checking the constraints and removing the tasks that do not satisfy the constraints, and does not take into account the case of adding tasks to the sequence, so the design difficulty and computational complexity of the algorithm are usually low. In order to facilitate the subsequent research, it is also necessary to introduce another special kind of task sequence: if the composition of the elements of a task sequence is composed by removing some elements from another task sequence, and the order relationship between the elements of this sequence is the same as that in the task sequence, then this sequence is called a subsequence of the corresponding task sequence.

4.2.2 Imaging quality analysis

In actual engineering projects, ensuring that the imaging satellite's imaging quality meets the user's needs is also an important constraint that the satellite management and control departments need to consider when scheduling the satellites' tasks. The constraints can be designed to ensure that the imaging quality of the task at the moment of execution is not lower than the minimum required imaging quality, which can be expressed by inequality (4.3):

$$MinQuality(i) \leq Quality(es_i) \quad \forall i \in \textbf{TS} \tag{4.3}$$

According to the description in [100], the imaging quality of an arbitrary task i that starts to be executed at the moment of es_i is calculated by equation (4.4):

$$\text{Quality}(es_i) = \left\lfloor 10 - 9\frac{|es_i - es_i^*|}{es_i^* - ws_i} \right\rfloor \quad \text{Iff } es_i \neq \text{null} \tag{4.4}$$

where es_i^* represents the time of the task start when the task i obtains the best image quality, and for a wide range of imaging sensors, the es_i^* corresponds to the moment when the absolute value of the pitch angle is the smallest. Therefore, es_i^* can be regarded as a constant after pretreatment of task i. In this way, equation (4.4) can be equivalently converted to the inequality (4.5):

$$\left\lceil es_i^* - |es_i^* - ws_i|\frac{10 - \text{MinQuality}_i}{9} \right\rceil \leq es_i \leq \left\lfloor es_i^* + |es_i^* - ws_i|\frac{10 - \text{MinQuality}_i}{9} \right\rfloor \tag{4.5}$$

Associative task visibility constraints, the start time of a successfully scheduled task es_i should satisfy inequality (4.6) and inequality (4.7):

$$\max\left\{ws_i, \left\lceil es_i^* - |es_i^* - ws_i|\frac{10 - \text{MinQuality}_i}{9} \right\rceil\right\} \leq es_i \tag{4.6}$$

$$es_i \leq \min\left\{we_i - d_i, \left\lfloor es_i^* + |es_i^* - ws_i|\frac{10 - \text{MinQuality}_i}{9} \right\rfloor\right\} \tag{4.7}$$

4.2.3 Analysis of attitude transition time constraints

Attitude transition time constraint is an important feature of agile imaging satellites. With the gradual increase in the refinement of satellite design, time-dependent agile imaging satellites have become the mainstay of today's remote sensing satellite field. "Time-dependence" means that the value of a variable changes with the change of a certain time variable. Time-dependent attitude transition time means that the value of attitude transition time between two consecutive tasks is related to the specific start and end imaging times of these two tasks. In this class of problems, the attitude transition time is then usually described as a function of these two task correlations. In this problem, the minimum time-dependent transition time is defined as a segmented function with respect to the synthesized angle of attitude maneuver, in conjunction with the definition of the time-dependent attitude transition time computation function in [100], as well as in real engineering projects, as detailed in equation (4.8):

$$\text{Trans}(i,j) = \begin{cases} 10 + \theta_{ij}/1.5 & \theta_{ij} \subset (0, 15] \\ 15 + \theta_{ij}/2 & \theta_{ij} \subset (15, 40] \\ 20 + \theta_{ij}/2.5 & \theta_{ij} \subset (40, 90] \\ 25 + \theta_{ij}/3 & \theta_{ij} \subset (90, 180] \end{cases} \tag{4.8}$$

where θ_{ij} denotes the synthetic attitude maneuver angle between task i and task j, which is calculated by combining the pointing angle at the end moment of task i with the pointing angle of the beginning of task j according to equation (4.9). The pointing angle of the task is denoted using a vector about time $(r_i(t), p_i(t))$:

$$\theta_{ij} = |r_i(t) - r_j(t)| + |p_i(t) - p_j(t)| \qquad (4.9)$$

It follows that the minimum attitude transition time is a function of the specific execution time of the tasks i and j.

4.2.4 Constraint checking algorithm based on timeline advancement mechanism

The actual imaging satellite task scheduling process usually contains a large number of complex constraints, which are one of the main factors contributing to the complexity of the problem. In order to reduce the degree of coupling between the constraint checking process and the main process of the scheduling algorithm, this book designs a constraint checking algorithm based on the timeline advancement mechanism. The algorithm is an independent function that can check whether the final scheme of the problem or the scheduling scheme generated by the intermediate process of the algorithm satisfies the constraints. This section first introduces the handling of the four classes of constraints in the algorithm, and then presents the overall solution framework of the algorithm.

(1) Cumulative constraints are constraints on physical quantities from the beginning to the end of a scheduling cycle, and the specific form of the constraints is described in Subsection 3.3.4 of this book. In this algorithm, an array Stat_k is used to record the cumulative constraint variables at any moment in the scheme to be checked, where $k = 1, 2, \ldots, k_1$ represents the number of the k for the first k cumulative constraint, and k_1 is the total number of entries of the cumulative constraints. These physical quantities are monotonically nondecreasing functions of time, i.e., $t_1 < t_2$, $\text{Stat}_k[t_1] \leq \text{Stat}_k[t_2]$ holds for any k, t_1, t_2. From the monotonicity of the cumulative constraints, the following theorem follows.

Theorem 4.1. *In the sequence corresponding to a task scheduling scheme, if there exists a subsequence constituting a task scheduling scheme that does not satisfy the cumulative constraint k, then the original task scheduling scheme does not satisfy the same cumulative constraint.*

According to Theorem 4.1, we can design the cumulative constraint checking process based on the timeline advancement mechanism, and the general idea is formulated as follows: the tasks in the scheme to be checked are listed in the order of their execution time, and then we add the corresponding cumulative constraint variables one by one. If the value of the variable exceeds the value required by the constraint after adding the tasks, then the overall scheme does not satisfy the cumulative constraint.

(2) Rolling constraints have similar properties to cumulative constraints in that they count the increments of specific physical quantities during a rolling period. In this algorithm, an array $Roll_k$ is used to record the rolling constraint variables at any moment in the scheme to be examined, where $k = 1, 2, \dots, k_2$ stands for the first k cumulative constraint, and k_2 is the total number of entries of the rolling constraints. According to the analysis of rolling constraints, it is found that the rolling constraint variables are also monotonically nondecreasing in the rolling period, so a calculation method similar to that of cumulative constraints can be designed: the tasks in the scheme are listed in the order of execution time, and then the tasks are added one by one and the corresponding rolling constraint variables are calculated, and if the difference between the value of any of the variables after the addition of the tasks and the value of the variable at the beginning time of the rolling period exceeds the corresponding value required by the rolling constraints. Then we can judge that the scheme as a whole does not satisfy the rolling constraints.

In particular, if there are constraint terms that denote the same physical meaning, such as recording the maximum number of attitude maneuvers of the day and the maximum number of maneuvers of the current orbiting cycle, the maximum cumulative imaging hours of the day and the maximum cumulative imaging hours of the current orbiting cycle an array is shared to record the corresponding values, and when checking the constraints, different judgmental statements are used to determine the rolling constraints or cumulative constraints. Then the time and space complexity of the algorithm can be effectively reduced.

(3) Task attribute constraints are the most challenging constraints to be described uniformly in mathematical form. However, some of the constraints can be merged by pretreatment, for example, the imaging quality constraint and visibility constraint are merged into one constraint in Subsection 4.2.2 of this chapter, and the constraints "imaging task cannot be performed during the time window of Earth shadow" and "imaging task cannot be performed in latitude higher than 60 degrees" can also be merged in a similar way. After merging some of the constraints, the constraint checking process of the task attribute constraints can be completed by checking all the task attribute constraints one by one. The time complexity of this process is $O(n)$.

(4) Task correlation constraints are based on the logical determination between two consecutive tasks, e. g., the determination of the minimum time interval between different types of tasks usually needs to be decided together with relevant operations, such as the attitude transition time calculation function, the minimum precession time, the minimum post-cession time, and other parameters of each task. Similar to the task attribute constraints, the partial correlation constraints can be merged for consideration, and it is sufficient to set the values that satisfy multiple constraints at the same time. After merging the partial constraints, the task correlation constraints are checked one by one, and the constraint checking process of the task attribute constraints can be completed. The time complexity of this process is $O(n)$.

By integrating the checking process of all the above constraints, this book proposes a constraint checking algorithm based on the timeline advancement mechanism, and the pseudocode can be found in Algorithm 4.1.

Algorithm 4.1 Constraint checking algorithm based on timeline advancement mechanism.

Input: Task scheduling scheme

Output: Constraint checking judgment results

1: Parameter initialization: r_0 = rolling period, Cumu_{k_1} [], Roll_{k_2} [], Att_{k_3}, and Corr_{k_4}.

2: **Temp** = The scheduling scheme after sorting in order of execution.

3: $i = 0$;

4: **while Temp** is not empty and $i <$ the number of tasks in the scheme **do**

5:     Read the first task **Temp**[i] in **Temp**;.

6:     $t =$ **Temp**[i].ExecuteEndTime;

7:     Calculate the attributes associated with the task **Temp**[i]: amount of storage consumed, attitude transition time, etc.;

8:     Computes the cumulative constraint statistic $\mathrm{Cumu}_{k_1}[t]$;

9:     **if** Existence of any k such that $\mathrm{Cumu}_k[t] - \mathrm{Cumu}_k[0]$ does not satisfy the constraint **then**

10:         **return** False;

11:     **end if**

12:     Computes the rolling constraint statistic $\mathrm{Roll}_{k_2}[t]$;

13:     **if** Existence of any k such that $\mathrm{Roll}_k[t] - \mathrm{Roll}_k[t - r_0]$ does not satisfy the constraint **then**

14:         **return** False;

15:     **end if**

16:     **if** Existence of a task attribute conflicting with a constraint value in Att_{k_3} **then**

17:         **return** False;

18:     **end if**

19:     **if** Existence of a task such that its correlation parameter with task i does not satisfy the setting in Corr_{k_4} **then**

20:         **return** False;

21:     **end if**

22:     $i = i + 1$;

23: **end while**

24: **return** TRUE;

Algorithm 4.1 implements the checking of all the constraints in the loop: lines 8–11 implement the checking of cumulative constraints, lines 12–15 implement the checking of rolling constraints, lines 16–18 implement the checking of task attribute constraints, and lines 19–21 implement the checking of task correlation constraints. Assuming that the number of tasks in the scheme of the input algorithm is n, the length of the task scheduling period is L, the number of rolling constraints variables and cumulative constraints variables entries is k_s, and the number of task attribute constraints variables and task correlation constraints variables entries is k_a. Then the number of program executions of the constraint checking algorithm count (the number of program executions of the constraint checking algorithm counts both numerical and logical operations, in which the sort algorithm adopts counting sort and the number of basic operations is $3L + n$.) can be estimated as

$$\text{count} = 3L + n + 1 + n(3 + k_a + Lk_s + k_a + k_s)$$
$$= 3L + 1 + n(4 + 2k_a + Lk_s + k_s) \tag{4.10}$$

The number of floating-point operations of the algorithm cannot be calculated exactly because the formulas used for each operation statement are different. Because the characteristics of the task scheduling scheme are also difficult to summarize, it is impossible to find the average number of operations. According to equation (4.10), it can be obtained that the time complexity of this constraint checking algorithm is the same as the time complexity of n, k_a, k_s, L are linear. It is worth noting that k_a, k_s, L do not change with the increase of task size n. Therefore, when the problem background is determined, k_a, k_s, L can be regarded as constants, and the time complexity of the constraint checking algorithm based on the timeline advancement mechanism is $O(n)$.

4.3 Heuristic algorithm based on the density of residual tasks

Efficient heuristic algorithms for solving complex scheduling problems usually consume minimal computational resources to achieve good solution results. However, constructing a reasonable heuristic function, which is the core of heuristic algorithms, usually requires deep theoretical skills and an in-depth understanding of the problem characteristics. Especially for complex scheduling problems, it becomes tough to guarantee the optimality of the algorithm. Combined with the above description of the problem model, this section designs a heuristic algorithm based on the residual task density after fully considering the characteristics of the task scheduling problem.

4.3.1 Thinking of the solution

For practical considerations, the constraints that need to be considered in the task scheduling process of imaging satellites are multidimensional: payload temperature constraints [94], platform and payload state constraints [96, 97], power constraints, solid-state memory constraints, and time-related constraints. However, by investigating the actual engineering problems, it is found that in the actual satellite operation control process, it is difficult to directly model the complex conditions existing in the real world and add them to the task scheduling model. For example, the payload temperature constraints are related to the operating conditions of the platform and payload, the satellite flight orbit, cosmic radiation, and many other complex conditions, which make it difficult to simulate this process with high accuracy. Therefore, this class of constraints is usually transformed into cumulative constraints, roll constraints, etc., like those in Subsection 3.3.4 of this book in order to ensure the regular operation of the satellite hardware. Most of these constraints are related to the time of execution of the imaging satellite task, and as the capability of the satellite hardware improves, the conditions related to the hardware capability, such as solid-state memory, will become more and more relaxed, and will not even be a constraint on satellite task scheduling in the future. It can be said that "time" is the most critical attribute in the process of satellite task scheduling, and it will become increasingly important in the future.

Therefore, in the task scheduling process, the decision maker usually wants to maximize the time utilization of the selected task when choosing a task, so that the tasks remaining with observation opportunities can be provided with the most generous scheduling time possible under the condition of choosing the same task, in order to enhance the probability of successful task scheduling for tasks that have not yet been considered. The heuristic algorithm HADRT based on the density of remaining tasks is born.

The flowchart of the heuristic algorithm based on the density of residual tasks (HADRT) is shown in Figure 4.4.

The algorithm receives the task information and resource information obtained from the task pretreatment, for practical considerations, the task information includes the profit of completing the task, the imaging time window, the visible time window, and the pointing angle at each moment during the visible time (each pointing angle is represented by a two-dimensional vector composed of the pitch angle and the roll angle), and the resource information includes the resource's capability parameter and the related computational functions, such as the attitude transformation function, the imaging quality functions, such as attitude conversion function, imaging quality function, etc. Then the process parameters are initialized, mainly setting the scheduling period, environment configuration parameters, etc. Then the tasks are sorted by the designed residual task density metric function, and based on this order, the information of the first task of the sequence is read and processed in combination with the information of the set of tasks that have been added to the scheduling scheme, and all constraints are taken into account to try to make a decision about the task. If the constraints are satisfied, the scheduling scheme is updated. Each time a decision on a task is completed, the task is removed from the sequence until all tasks in the sequence have been considered.

The basic idea of the algorithm is to first obtain a sequence of tasks based on the constructive heuristic function, and then make a decision on the task start time based on the sequence. It can be seen that the process of ordering the tasks based on the residual

```
                              ┌──────────────┐
                              │    Start     │
                              └──────┬───────┘
                                     ▼
┌──────────────────┐         ╱─────────────────╲
│ Task pretreatment │ ──────▶│    Input data    │
└──────────────────┘         ╲─────────────────╱
                                     ▼
┌──────────────────┐         ┌─────────────────┐
│   Environment    │         │  Initialization │
│    Resource      │         └────────┬────────┘
│  Orbital elements│                  ▼
└──────────────────┘         ┌─────────────────┐
                             │ Sort tasks based on the │
                             │ remaining task density  │
                             └────────┬────────┘
                                      ▼
                             ┌─────────────────────┐
                             │ Read and process the │
                             │ first task in the sequence │
                             └────────┬────────┘
                                      ▼
                             ┌─────────────────────┐
                             │ Try to determine the │
                             │    imaging time      │
                             └────────┬────────┘
```

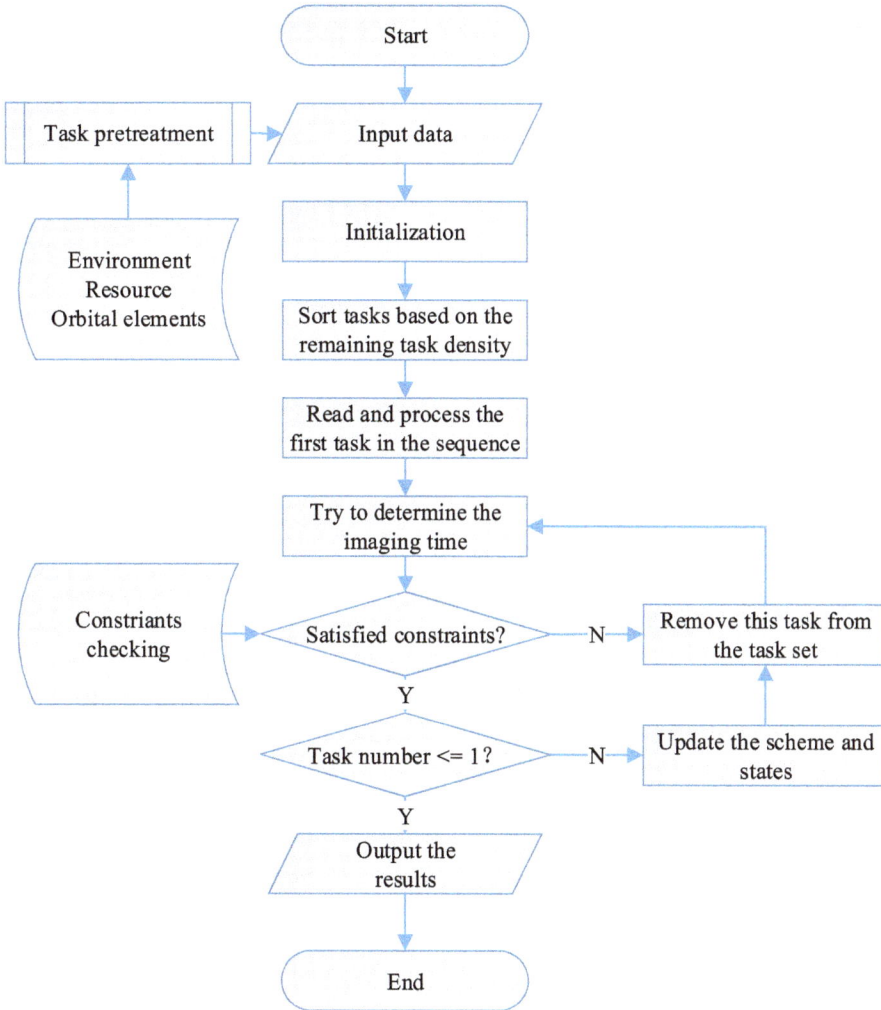

Figure 4.4: Framework of HADRT.

- Task pretreatment
- Input data
- Environment / Resource / Orbital elements
- Initialization
- Sort tasks based on the remaining task density
- Read and process the first task in the sequence
- Try to determine the imaging time
- Constriants checking — Satisfied constraints? — N → Remove this task from the task set
- Y
- Task number <= 1? — N → Update the scheme and states
- Y
- Output the results
- End

task density plays a significant role in the solution efficiency of the whole algorithm. In the next section, this constructive heuristic function is designed and analyzed in detail.

4.3.2 Constructing the heuristic function

The core idea of the heuristic function based on residual task density is to seek a high utilization of task scheduling schemes in the time dimension in order to improve the completion rate and overall profit of the scheduled tasks. It can usually be assumed that all other conditions being equal, if the selection of one task at a time leads to the highest density of desired tasks that can be scheduled in the future, the greater the number of these tasks that are successfully scheduled and the greater the expected total profit of the final scheme.

First, a function is defined to compute the density of the distribution of tasks in terms of time: the ratio of the total available scheduling time windows to the average task occupancy time. Where the available scheduling duration can be expressed as the difference between the end moment of the scheduling period and the end time of the selected task ee_i, a task occupancy time is the sum of a task's imaging duration and the attitude transition time between two tasks, which can be expressed as the ratio of $d_i + \mathrm{Trans}(i-1, j)$ to calculate the average task occupancy time, i. e., the average of the occupancy times of all the remaining candidate tasks under a certain decision condition. According to this definition, accurately finding the density of remaining tasks under all possible conditions is very time consuming, contrary to the method design's original intention. Based on the reality, the above function is simplified, and this book constructs a heuristic function based on the density of remaining tasks:

(1) The scheduling period is specified during the problem modeling process, so it can be viewed as a constant. Discussing the ratio of the available scheduling duration to the average task occupancy is equivalent to discussing the ratio of the end moment of the selected task ee_i to the average task occupancy.

(2) Regarding the attitude transition time function, it is usually an axisymmetric function about the synthetic attitude maneuver angle. According to the analysis of [133], if the task is randomly distributed, the attitude transition time function has its expectation value as a summable constant. Therefore, replacing the actual attitude transition time for every two neighboring tasks with the expected attitude transition time, $\overline{\mathrm{Trans}}$ simplifies the computation and facilitates the subsequent analysis of the method.

Based on the above discussion, the heuristic function based on the density of residual tasks is constructed as follows:

$$H(i) = \sum_{k \in \mathbf{HQ}} \frac{-ee_i}{\frac{1}{\mathrm{card}(\mathbf{HQ})}(d_k + \overline{\mathrm{Trans}})}$$
$$= \sum_{k \in \mathbf{HQ}} \frac{-ee_i(\mathrm{card}(\mathbf{HQ}))}{d_k + \overline{\mathrm{Trans}}} \tag{4.11}$$

included among these

$$\mathbf{HQ} = \{k \in \mathbf{TSQ} \mid ee_k > ee_i\} \tag{4.12}$$

In particular, $\mathbf{HQ} = \phi$, define $H(i) = 0$. In equation (4.11), the greater numerator $-ee_i$ is, the earlier the visible time window of the task i ends. It is positively correlated with the available scheduling duration. The denominator represents the average task occupancy of all tasks that execute later than task i. Therefore, the algorithm selects the task that maximizes the function $H(i)$, i. e., arg max $H(i)$. To determine whether or not to accept a task i, a decision on the task start time and end time for that task is also required. The algorithm uses a tight scheduling strategy to determine the start time for the task: i. e., the earliest observation moment point is selected to execute the task, provided that all the constraints are satisfied. The method for determining the start time of the task is as equation (4.13):

$$es_i = \max\{es_i^*, ee_{i-1} + Trans_{i-1,i}(es_{i-1}, es_i)\} \qquad (4.13)$$

The schematic of the two key steps of the algorithm (heuristic strategy for ordering tasks, and tightly strategy for task start time decision) is shown in Figure 4.5.

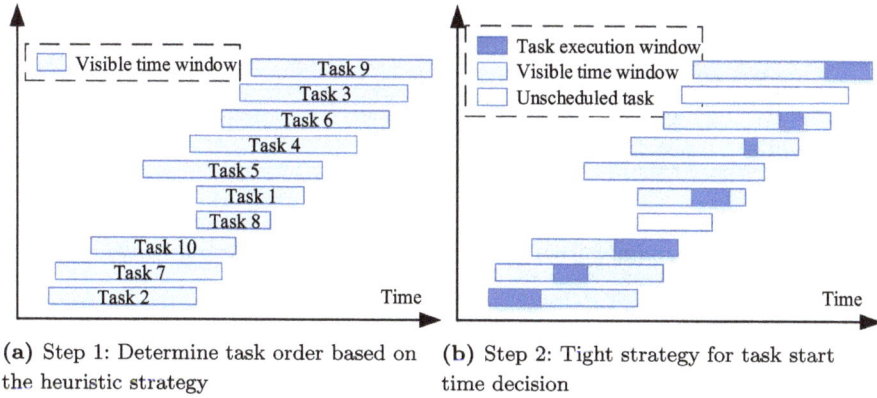

(a) Step 1: Determine task order based on the heuristic strategy

(b) Step 2: Tight strategy for task start time decision

Figure 4.5: Two key steps of HADRT.

Thus, the pseudocode for HADRT is shown in Algorithm 4.2.

4.3.3 Proof of optimality

Concerning the above constructive heuristic algorithm, there is the following theorem.

Theorem 4.2. *When the "user's requirement for imaging" is regarded as a random event, in addition to meeting the basic requirements of the task scheduling problem, the total expected benefit of the solution obtained by HADRT is maximized as long as the following five assumptions are met:*
(1) *The attitude transition time between tasks is constant.*
(2) *The imaging duration follows a specific distribution whose expected value can be found.*
(3) *The task benefits follow a particular distribution whose expected value can be found.*

Algorithm 4.2 Heuristic algorithm based on the density of residual tasks (HADRT).

Input: Set of decided tasks **TS**,

Output: Scheduling scheme

  1: Parameter initialization;

  2: **while** Task sequence is not empty **do**

  3:    **for** $i = 1$ **to** |TSQ| **do**

  4:       Tag $\leftarrow 0$;

  5:       Calculate $H(i)$ of the task i;

  6:       **if** $H(i) > $ Tag **then**

  7:         Tag $\leftarrow H(i)$;

  8:       **end if**

  9:    **end for**

10:    Use tight strategies to set task start times for tasks;

11:    **if** Adding task i without violating constraints **then**

12:       Accepts task i to the scheme and removes task i from the task sequence **TSQ**;

13:       Update scheduling scheme;

14:    **end if**

15: **end while**

---

(4) *There is no entire containment relationship of the task's visible time window between tasks.*

(5) *Resource capacity constraints such as storage and power are not considered.*

Many practical engineering application scenarios basically satisfy these conditions. For example, in the process of attitude maneuvering of some models of satellites, no matter how the pointing angle changes, their attitude transition time is a constant value, which satisfies the assumption (1); when all the tasks correspond to point-target imaging, the imaging duration is usually a small value of constant, which satisfies the assumptions (2) and (4); and in the scenarios where the task profit is difficult to be metricized, it is usually considered that the importance of each task is not different, satisfying assumption (3). In practical applications, time-related constraints are usually tight constraints, and other constraints such as power, solid-state memory, and other resource capacity constraints are loose constraints, i. e., most satellites can be designed to ensure that the resource capacity constraints are not violated when maximizing the utilization of the tasks in the time dimension, satisfying assumption condition (5). Therefore, it is of strong practical guidance and theoretical significance to study the optimality of the method when these conditions are satisfied. By combining the mathematical induction method and the inverse method, it can be proved for Theorem 4.2.

Proof. Such that the solution obtained by HADRT is recorded as $S_{\text{HADRT}} = \{i_1, i_2, \ldots, i_m\}$, assuming that there exists an optimal solution to the problem $S_{\text{opt}} = \{j_1, j_2, \ldots, j_n\}$. Assuming that the tasks in both sets are in time order and there is no time window overlap of tasks in the scheduling scheme, there are

$$\forall k < n, \quad ee_{i_k} \leq es_{i_{k+1}} \tag{4.14}$$

$$\forall k < m, \quad ee_{j_k} \leq es_{j_{k+1}} \tag{4.15}$$

To prove that the expected profit of the solution obtained by the HADRT is maximized, it is only necessary to prove the expected profit of the solution obtained by HADRT $E(S_{HADRT})$ equals the expected profit of the optimal solution $E(S_{opt})$; to prove that $E(S_{HADRT}) = E(S_{opt})$, it is only necessary to prove that the scheme that the number of successfully scheduled desired tasks in S_{HADRT} is equal to that in $S_{opt}m$ by combining with assumption (3), i. e., $n = m$. The following two lemmas are given.

Lemma 4.1. *The execution end time ee_{i_k} of any of the tasks i_k in the solution obtained by HADRT is not greater than the execution end time ee_{j_k} of the task j_k in the corresponding position in the optimal solution, in which $(1 \leq k \leq m)$.*

Proof (By mathematical induction).
① When $n = 1$, select task according to equation (4.11) and equation (4.12)

$$i_1 = \arg\max_{i_1} H(i_1) = \arg\max_i \sum_{k \in TSQ} \frac{-ee_i(card(\mathbf{TSQ}))}{d_k + \overline{Trans}} \tag{4.16}$$

According to assumption (1) and assumption (2), there are

$$E\left(\frac{d_{i_1} + \overline{Trans}}{card(\mathbf{TSQ})}\right) = \frac{\overline{d} + \overline{Trans}}{card(\mathbf{TSQ})} \tag{4.17}$$

In this condition, the expected values of most variables are constants, and the only variable that affects task selection is the time window end time ee_{i_1} of the candidate tasks, as in equation (4.18), so choosing the task with the earlier end of the time window in the set of tasks to be selected each time ensures that the desired residual density is maximized:

$$\overline{H(i)} \propto h_1(i) = -ee_i \tag{4.18}$$

According to the algorithmic idea of HADRT, the first task to join the scheme should be the one corresponding to the earliest visible time window end time among all the tasks, and due to the assumption of condition (4), there is a task i_1 corresponding to the start time of the visible time window which is also the smallest among all the tasks, and since there are no other tasks in conflict, in accordance with the tight strategy, its actual task start time is the start time of the visible time window. Combining all the above conditions, there are

$$es_{i_1} \leq es_{j_1} \tag{4.19}$$

② Assume that Lemma 4.1 holds when $x = k(1 \le k < m)$. It can be inferred that

$$es_{i_k} \le es_{j_k}. \tag{4.20}$$

When $x = k+1$, based on the assumption (3) and the nature of the optimal solution, the number of successfully scheduled tasks in S_{opt} should be no less than all other solutions, i. e., $m \le n$. So, any one of the tasks in S_{HADRT} with subscript x $(1 < x \le m)$ can be found to correspond to it, and the task j_x has the following properties:

$$ee_{j_{x-1}} \le es_{j_x}, \tag{4.21}$$

and because inequality (4.20) holds, it can be concluded that

$$ee_{i_{x-1}} = ee_{i_k} \le ee_{j_k} \le es_{j_{k+1}} = es_{j_x}. \tag{4.22}$$

Obviously, the imaging end time of task j_x of S_{opt} lies in the interval $[ee_{i_{x-1}}, +\infty]$ and according to HADRT, the task i_x in solution S_{HADRT} is the task corresponding to the earliest window in this interval, i. e., $ee_{i_x} \le ee_{j_x}$ holds.

In summary, for any natural number $k \in [1, m]$, $ee_{i_k} \le ee_{j_k}$ holds. Lemma 4.1 is proved.

□

Lemma 4.2. *The number of tasks in S_{opt} equals the number of tasks in S_{HADRT}, that is,* $m = n$.

Proof (By reduction to absurdity). Assuming $m \ne n$, according to the condition "the number of tasks in the optimal solution S_{opt} is not less than the number of tasks in all other solutions," there are

$$m \le n. \tag{4.23}$$

Thus, there exists task $j_k(k > m)$ that starts after task j_{k-1}. According to Lemma 4.1, there are

$$ee_{i_{k-1}} \le ee_{j_{k-1}}, \tag{4.24}$$

i. e., task j_k also starts after task i_{k-1}, and task j_k is added to solution S_{HADRT} in which the constraints are not violated.

According to the design of Algorithm 4.2, the algorithm ends with the condition that the sequence of tasks is empty, and in combination with assumption (5), the task j_k can be added to the solution S_{HADRT}.

At this point, the number of tasks in the solution S_{HADRT} is $m + 1$, which is contradictory to the precondition.

Therefore, $m \ne n$ is not true, i. e., $m = n$ is proved. □

By the proofs of Lemma 4.1 and Lemma 4.2, the proof process of Theorem 4.2 is a success. □

4.3.4 Complexity analysis

In terms of time complexity, the main process of Algorithm 4.2 consists of two layers of loops: the outer loop is the one that implements the traversal of tasks and selects the tasks based on the construction of the heuristic function and updates the scheme, and the inner loop computes each task heuristic function for each task. Since the outer loop has the same conditions as the outer loop of the constraint checking algorithm, the nested constraint checking process can be realized by calling Algorithm 4.1 in line 11 only from line 5 to line 22. Assuming that the number of tasks in the scheme of the input algorithm is n, the average number of executions of program statements for this constraint checking algorithm can be estimated by equation (4.25):

$$
\begin{aligned}
count &= n + \left(\frac{1}{2} \times 4 + \frac{1}{2} \times 3 \right) \times (n + n - 1 + \cdots + 1) + n \times \left(1 + \frac{3}{2} + \frac{2}{2} \right) \\
&= \frac{7}{4} n^2 + \frac{29}{4} n
\end{aligned}
\tag{4.25}
$$

The above computation process in the judgment statement is recorded as an average of 0.5 operations. Since the complexity of Algorithm 4.1 lines 5 to 22 is $O(1)$, the combined time complexity of HADRT is $O(n^2)$ according to equation (4.25).

In terms of space complexity, the program implementation of HADRT requires only two structural variables for storing the initial task information and the scheduling scheme information, two integer variables as the loop parameters and label parameters, respectively, and not recursive in the program. The dynamic memory size occupied by the program grows linearly with the number of tasks, so the space complexity of HADRT is $O(n)$. In simulation experiments, running HADRT requires less than 10 kB of memory when the number of tasks does not exceed 2000.

4.3.5 Analysis of strengths and limitations

By analyzing Theorem 4.2, it is demonstrated that the method can obtain a high-quality solution to the imaging satellite scheduling problem at a small cost in time and space. In addition, since the algorithm treats each visible time window of each task as an independent object during the computation process, HADRT can also be used to solve the imaging satellite task planning problem without adjusting the algorithm structure directly. However, the limitations of the algorithm in practical application are apparent, which are mainly reflected in the following three points:
(1) Maximizing the expected profit does not mean that every scheduling process will result in a satisfactory solution. In the process of calculating the expected total profit, all variables are calculated using their expected values instead. Some of the variables have a scattered distribution of values, and the real values may differ too much from their expected values, resulting in the actual profit corresponding to the

scheduling scheme constructed on the basis of the real values being smaller than the expected profit.

(2) The "specific distribution" mentioned in the assumptions, however, in the actual scheduling process, it is difficult to find a specific distribution to describe the statistical pattern of the corresponding variable, so the corresponding expected value is hard to find.

(3) It is difficult to theoretically analyze the gap between the quality of the solution and the optimal solution of HADRT for the general imaging satellite scheduling scenarios.

The experiments in this chapter allow for the verification of the solution effectiveness and computational efficiency in a relatively generalized scenario.

4.4 Dynamic programming algorithm based on task sequencing

As an important branch of operations research, dynamic programming is one of the popular methods for solving optimization problems [35]. It specializes in solving multistage decision-making problems with stage determination. In terms of the nature of the object of study, multistage decision-making problems with discrete decision variables and combinatorial optimization problems are both problems that seek the extreme value of an objective function under multiple constraints [17]. In this section, the task scheduling problem is considered to be divided into two steps, and in the first step, a heuristic algorithm is used to realize the sequencing of tasks, and based on the task sequence, the problem can then be regarded as a multistage decision-making problem, which in turn can be used to use a dynamic programming algorithm based on task sequencing.

4.4.1 Multistage decision-making model

Under the condition that the task sequence is determined, the imaging satellite task scheduling problem is described as a multistage decision-making model and solved by a dynamic programming algorithm that must follow the principles of the model [134, 135] in multistage decision-making theory, and the following elements are clarified:

(1) Stage
In the imaging satellite task scheduling problem, if the task sequence can be obtained, the execution order of the tasks can be considered as the natural basis for the stage division: the decision process of the task i in the task sequence can be considered as the stage i of the multistage decision-making process, and the dynamic programming is solved according to the order of the stages in this stage. It is worth noting that the "stages" here are for the multistage decision-making model, and the states, decisions,

strategies, etc. in each stage can be described and analyzed in a unified way, and the decision-making problem in each stage can be described in the same form.

(2) Status

The state records a vector of variables in the model that affect decisions and strategies. Conventionally, a state variable of stage i is denoted by x_i, which consists of several terms $q_0, q_1, \ldots$. In the task scheduling problem, q_0 is defined as the objective function value obtained by the state x_i (which is generally the total profit obtained before the current state), q_1 records the state x_i is at the moment point, q_2 records the antecedent tasks in that state, and the subsequent terms are constraint-related terms.

The set of feasible states comprising all possible states in this phase is denoted by X_i [134]. In the multistage decision-making model based on the scheduling problem described in Subsection 3.4.2 of this book, the state variables consist of at least the attributes of the moment in time of the state *CurrentT*, the number of the previous task *LastTask*, the remaining available imaging time *RemainTime*, and the remaining available solid-state memory *RemainStore*. The set of allowed states, i. e., the set of all possible values of this vector is described. The state variables and the set of feasible states for the phase i are shown in equation (4.26) and equation (4.27).

$$
\begin{aligned}
x_i = {} & (\text{CurrentT}_1, \text{LastTask}_1, \text{RemainTime}_1, \text{RemainStore}_1) \\
& \text{or } (\text{CurrentT}_2, \text{LastTask}_2, \text{RemainTime}_2, \text{RemainStore}_2) \qquad (4.26) \\
& \text{or} \ldots
\end{aligned}
$$

$$
\begin{aligned}
X_i = {} & \{(\text{CurrentT}_1, \text{LastTask}_1, \text{RemainTime}_1, \text{RemainStore}_1), \\
& (\text{CurrentT}_2, \text{LastTask}_2, \text{RemainTime}_2, \text{RemainStore}_2), \qquad (4.27) \\
& \ldots\}
\end{aligned}
$$

The following points need to be made about the states in this multistage decision-making model:

(1) The states in the model characterize the task scheduling problem and are without an aftereffect. No aftereffect is defined as Definition 4.3 [134]. Statistical variables are recorded during the state design process, and the state at any given stage i can characterize the impact of decisions and strategies taken from stage 1 to stage i, and based on the state of each stage, it is possible to backtrack to find the decisions made at each previous stage.

Definition 4.3 (No aftereffect [134]). Once the state of a stage i has been determined, the states of all subsequent stages can be obtained by combining the decision of the state variables of the stage i, independent of all previous states.

(2) All time variables in the problem need to be discretized. The task scheduling problem is accustomed to discretizing the time variables in 1-second intervals, i. e., all

calculations do not retain decimals, and the results of the operations of all time-related parameters are eventually taken as integers. Therefore, using a multistage decision-making model to describe the decision-making process of the task start time in the imaging satellite scheduling problem is reasonable.

(3) State variables are an essential basis for checking constraints. For different practical problems, the state variables need to be adjusted accordingly due to different constraint considerations. For example, if there is a constraint that "the cumulative duration of imaging time per day should not exceed XX seconds" in a specific engineering project, then one item should be added to the state variable: cumulative duration of imaging time. This item will be updated as the state transitions, and the relevant constraints will be checked. Therefore, the dimensionality of the state variable in the model grows linearly with the number of constraints. The constraint terms in the imaging satellite scheduling problem generally do not grow explosively, so the dimensionality of the state variables in this method does not exceed an acceptable range.

(3) Decision-making
If certain choices can be made to change the state, these choices are called decisions. The decision in this problem is "whether to start the task i at time t." Therefore, assuming that the length of the visible time window of the first i task in a task sequence is 200 s and the imaging time is 10 s, based on the assumption of discretization of the variables, there are 191 state variables corresponding to the decision-making, i. e., every second from the beginning of the visible time window (denoted as 0s) to the end of the imaging time, i. e., the 190th second of the visible time window. The algorithm needs to make a decision on whether to perform the task in each state. It is customary to use $u_i(x_i)$ to represent the decision variables for the state x_i at the stage i, $U_i(x_i)$ represents the set of allowed decisions [134]. For the imaging satellite task scheduling problem, the decision variables and the set of allowed decisions are shown in equation (4.28) and equation (4.29), respectively.

$$u_i(x_i) = \begin{cases} 0, & \text{in state } x_i \text{ when shedding task } i+1 \\ 1, & \text{in state } x_i \text{ when accepting task } i+1 \end{cases} \tag{4.28}$$

$$\forall x_i, \quad u_i(x_i) = \{0,1\} \tag{4.29}$$

where the attribute q_1 of the state x_i is the point in time at which the state is located, which should satisfy the time window constraint, i. e.,

$$x_i[q_1] \in \mathbb{N}, \quad x_i[q_1] \in [es_i, ee_i]. \tag{4.30}$$

(4) Strategy

A sequence of decisions is called a strategy. $p_{i,j}(x_i)$ denotes the strategy from the state x_i of stage i to the stage j. The relationship between strategy and decision is shown in equation (4.31):

$$p_{i,j}(x_i) = \{u_i(x_i), u_{i+1}(x_{i+1}), \ldots, u_j(x_j)\}. \tag{4.31}$$

There is more than one strategy between any two stages i and j. One particular strategy is known as the optimal strategy, i. e., the strategy corresponding to the optimal value of the objective function of the problem, denoted as $p_{i,j}^*(x_i)$.

(5) State transfer equation

By the nature of the multistage decision-making model, based on a state x_i and the decision $u_i(x_i)$ in this state, the next state x_{i+1} can be determined. The calculation of determining x_{i+1} by x_i and $u_i(x_i)$ is known as the state transition equation. The inverse of this process, i. e., based on a state x_i and a target state x_{i+1} to find the corresponding decision variable $u_i(x_i)$, can be used to solve the optimization problem. In the imaging satellite task scheduling problem, the state transition equations for each term in the state are related to changes in objective conditions, such as resource consumption, that are specifically considered. For example, the objective function determines the state transition of the q_0; the attitude transition time function determines the state transition of the state's moment point q_1 and the predecessor task q_2; and the solid-state memory consumption model and the attitude transition time model determine the corresponding state transition of the other terms.

Taking the state transition of the value of the objective function q_0 as an example, when the objective function of the task scheduling problem is the maximization of the sum of the profits, the state transition equation in this model is shown in Figure 4.6.

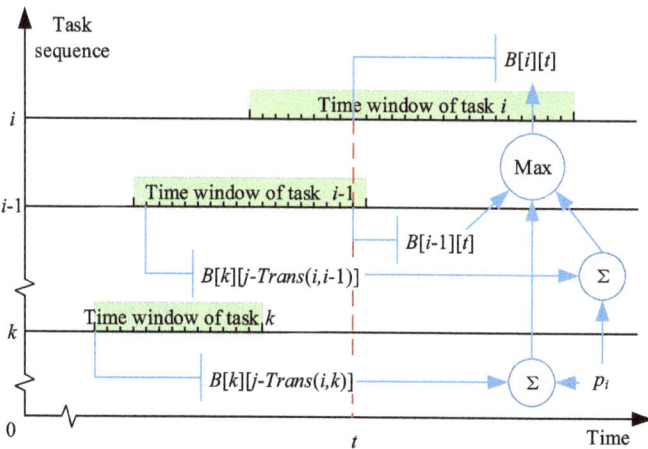

Figure 4.6: State transition in DPTS.

In the diagram, to simplify the labeling, $B[i][t]$ denotes the value of total profit at $x_i[q_1] = t$ in stage i. Schematically, the state at the t moment of stage i can be transitioned in two ways:

(1) When the task of stage i is discarded, the value of the objective function in the state of the current stage is equal to the value of the objective function in the state of the corresponding moment of the previous stage, i. e.,

$$B[i][t] = B[i - 1][t]. \tag{4.32}$$

(2) When accepting a task in stage i, the value of the objective function in the state of the current stage can be equal to the sum of the value of the objective function of the state corresponding to all the moments in the previous stage that satisfy the constraints and the profit of the task corresponding to the current state, i. e.,

$$B[i][t] = B[i - k][t - d_i - \text{Trans}(k, i)] + p_i \tag{4.33}$$

where d_i is the imaging duration of the task, $\text{Trans}(k, i)$ is the imaging duration of the tasks k and the pose transition time of task i, and p_i is the profit of task i. Whether this state transition process is feasible or not needs to check whether the constraints are violated after joining the task i, which is determined by trying to compute the state after the transition in conjunction with the feasible range of values of each item in the state.

The objective function is to maximize the cumulative profit, so the state transition equation is designed as follows:

$$B[i][j] = \max_k\{B[i - 1][j], B[i - k][j - d_i - \text{Trans}(k, i)] + p_i\} \tag{4.34}$$

Based on this equation, the decisions at each stage can be derived to obtain the solution strategy for the whole problem.

4.4.2 Main computational process

The pseudocode for the DPTS proposed in this book is detailed in Algorithm 4.3.

In Algorithm 4.3, lines 6 through 14 are the implementation of the state transition procedure that is the core of the DPTS. Since the outermost loop has the same looping conditions as the constraint checking procedure, the constraint checking procedure in line 8 is similarly implemented by calling lines 5 through 22 in Algorithm 4.1. The algorithm uses a matrix with the scale card(**TSQ**) $*$ (ee$_{\text{card(TSQ)}-1}$) to record the value of the optimal objective function for each state, and the solving process can be realized through a three-level loop.

Algorithm 4.3 DPTS.

Input: Task set **TS**

Output: Optimal value of the objective function $B[\text{card}(\textbf{TSQ})][ee_{\text{card(TSQ)}-1}]$

1: Variable initialization;

2: Constructs a sequence of tasks **TSQ** based on the ascending order of the latest end time of the candidate tasks;

3: scheduling interval $\leftarrow [ee_{\text{card(TSQ)}-1}]$;

4: **for** $i \in$ **TSQ do**

5:    **for** $t = es_0 : ee_{\text{card(TSQ)}-1}$ **do**

6:       **for** $k = 0 : i$ **do**

7:          Calculate the value of Temp $= B[i - k][t - d_i - \text{Trans}(k, i)] + p_i$.

8:          **if** Temp $> B[i][t]$ and do not violate constraints **then**

9:             $B[i][t] \leftarrow$ Temp.

10:          **else**

11:             $B[i][t] \leftarrow B[i - 1][t]$.

12:          **end if**

13:          k++;

14:       **end for**

15:       t++;

16:    **end for**

17: **end for**

4.4.3 Proof of optimality

The process of deciding a specific start time for each task, given that the task sequence is determined, and finding the scheduling result that maximizes the total profit of the scheme while satisfying all the constraints is a key step in solving the task scheduling problem. Regarding the multistage decision-making model and DPTS for the imaging satellite task scheduling problem, the following theorem holds.

Theorem 4.3. *In addition to satisfying the basic conditions of the task scheduling problem, the DPTS guarantees that the resulting scheme is optimal when the following three assumptions are satisfied:*

(1) *The task sequence corresponding to the global optimal solution is a subsequence of the constructed task sequence TSQ.*

(2) *At any stage i, it is possible to determine, based on the state of the current stage, how well the previous i stage, using a certain strategy $p_{i,j}(x_i)$ the constraint satisfaction of the scheme composed.*

(3) *The time variable can be discretized.*

The loop conditions of the constraint checking algorithm based on the timeline advancement mechanism (i. e., Algorithm 4.1) can be effectively merged with the intermediate level loops in the DPTS (i. e., Algorithm 4.3). Therefore, Algorithm 4.1 can be nested in Algorithm 4.3 to realize the constraint checking process, which satisfies the assumption (2), and the optimal solution of the task scheduling problem under the condition that the optimal task arrangement is known can be found using this method. To prove that the above theorem holds, it is only necessary to show that the dynamic programming part of the algorithm satisfies the principle of optimality [134] when solving the above multistage decision-making problem, i. e., Theorem 4.4.

Theorem 4.4. *A necessary and sufficient condition for a strategy $p_{1,n}^*(x_1)$ to be optimal is that for any $k \in (1, n]$ there are strategies $p_{1,k-1}^*(x_1)$ and $p_{k,n}^*(x_1)$ are able to obtain the optimal objective functions of the two processes, respectively.*

Theorem 4.4 is one of the basic theorems of dynamic programming, and its proof procedure is not repeated here. With the help of Theorem 4.4, Theorem 4.3 can be proved. The proof is as follows.

Proof (By reduction to absurdity). Suppose Algorithm 4.3 does not meet the principle of optimality in solving the above multistage decision-making problem, i. e., when $B[n][t_{max}]$ is maximized, there is a certain intermediate stage $k \in (1, n)$, such that the objective function value of the subprocess of the first half is $B[k-1][x_{k-1}[q_1]]$ or the objective function value of the subprocess of the second half is not the maximum objective function value of the pairwise process. The imaging satellite task scheduling problem defines the objective function as maximizing the total profit, so it is computed as the sum of the objective function values of the first half of the subprocesses and the objective function values of the second half of the subprocesses. Therefore, it is discussed in separate cases:

(1) If the objective function value of the first half of the subprocess is not the maximum function value, then there exists another strategy $p_{0,k}'(k_1)$ such that $B[k-1][x_{k-1}[q_1]]$ takes a larger value. Also, because the proposed multistage decision-making model has no aftereffect, the calculation of the objective function value for the subprocess of the second half is independent of the previously adopted decisions and strategies, and only refers to the state x_{k-1} at stage $k-1$. Therefore, if the strategy of the subprocess of the second half remains unchanged, the total objective function value obtained by solving the subprocess of the first half with the strategy $p_{0,k}'(k_1)$ is larger than the original objective function value $B[n][t_{max}]$, i. e., the original objective function value $B[n][t_{max}]$ is not optimal, which is contradictory to the assumption.

(2) Since the subprocess of the second half will have no effect on the computation of the first half, based on assumption (1), we discuss the decision-making problem starting from stage k. Obviously, if the objective function value of the subprocess of the second half is not the maximum function value, there exists another strategy to make the objective function value better, and there exists a better solution to the problem composed of the subprocess of the first half and the second half, which is contradictory to the assumption.

In summary, DPTS has the optimal substructure property in solving the above multi-stage decision-making problem. At the same time, the above multistage decision-making problem satisfies the nature of no aftereffect, so Theorem 4.3 is proved. □

4.4.4 Complexity analysis

In terms of time complexity, the main process of Algorithm 4.2 consists of three layers of loops, with the outermost loop realizing the traversal of the tasks, the intermediate loop where each task is traversed at each point in time from the beginning of the scheduling period to the end of the scheduling period, and the inner loop traverses the tasks before the current task, i. e., it can determine all constraints and realize state transition. Assuming that the number of tasks in the scheme of the input algorithm is n and the length of the task scheduling period is L, the average number of executions of the program statements of DPTS can be estimated by equation (4.35) to estimate it:

$$\text{count} = 2 + 3L + n + n * L * 4 * (1 + 2 + \cdots + n - 1)$$
$$= (n^3 - n^2 + 3)L + n + 2 \tag{4.35}$$

The judgment statement is recorded as 0.5 operations on average in the above computation. Since the complexity of lines 5 to 22 in Algorithm 4.1 is $O(1)$, according to equation (4.35), the combined time complexity of DPTS is $O(n^3)$.

Algorithm 4.3 needs to record the constraint values of rolling constraints and cumulative constraints in each state for updating the state. So, in the process of implementation, Algorithm 4.3 needs to create two structural variables for storing the initial task information and the scheduling scheme information, respectively: one structural variable for storing the state in the decision-making process, and three integer variables as the loop parameters and label parameters. In addition, $k_s + 1$ matrices of $n * L$ are needed to record intermediate values of k_s constraints as well as the values of the objective function in each phase and state. Although the total amount of memory required is much larger than the amount of Algorithm 4.2, the memory space occupied by the algorithm varies linearly with n, L, and k_s.

4.4.5 Analysis of strengths and limitations

The DPTS has more advantages in solving the task scheduling problem than other exact algorithms:
(1) Although the task scheduling problem is not an NP-hard problem under the condition of determining the task sequences, the solution space of this problem is still huge. The use of tree search and local search will always face the problems of inefficient search and easy cause a dimensional catastrophe, so it is difficult to ensure

that the global optimal solution of the problem can be obtained in polynomial time. The DPTS can ensure that the optimal solution of the problem in polynomial time. It maximizes the solution efficiency and the quality of the solution simultaneously.

(2) The solving process can be decoupled from the various constraints. The constraint checking algorithm proposed in Subsection 4.2.4 of this chapter is nested in the DPTS, and it is sufficient to reasonably design the variables related to the constraints and the algorithm of checking constraints, and check the constraints in the course of each state transition. Therefore, the algorithm's optimality do not change with the constraints.

(3) The operation time of the DPTS is stable and does not change with conditions such as the distribution characteristics of the task and various constrains, so the time required for the operation of the algorithm can be estimated based on the time complexity. Stable algorithms tend to be more widely used in engineering, and are more reliable in scenarios that require a precise grasp of the computation time.

Also, the algorithm has the following limitations:

(1) DPTS proposed in this book can only find the optimal solution of the scheduling scheme under a specific task sequence. However, there are $n!$ arrangements of a set with n tasks, and the time complexity of solving the task sequencing problem by using dynamic programming to divide the phases and unfold the different arrangements is $O(n!n^3)$. On the other hand, with the high space complexity of the dynamic programming algorithm, this approach may consume too many computational resources in practice, and it is difficult to obtain a solution to the problem in acceptable time [133].

(2) The number of variables that need to be defined by the DPTS increases linearly with the dimensionality of the state. The state of each variable at all time in the scheduling period needs to be recorded, and the space complexity increases with the constraint entries. Therefore, Algorithm 4.3 consumes much more memory than Algorithm 4.2. When used in integration with other algorithms or when there are constraints on computational resources, such as when combining machine learning that requires storing a large number of intermediate processes, or when scheduling is performed on satellite-carried on-board computers, the algorithms may encounter arithmetic bottlenecks when dealing with real-world problems with a large number of constraint terms.

4.5 Simulation experiment

In order to verify the effectiveness of HADRT and DPTS in the imaging satellite scheduling problem, the following simulation experiments are designed in this section. First, the details of the experimental scenario are introduced, and then the algorithms' solution accuracy and computational efficiency are discussed through the analysis of the algorithms' results in this scenario, and finally the relevant conclusions are obtained. All experiments are implemented and compared on a laptop with Intel Core i7-8750H CPU @ 2.20 GHz, 16 GB RAM, and NVIDIA GeForce GTX 1060.

4.5.1 Scenarios design

(1) Satellite parameters and capability

1) *Design of satellite orbital elements*

The orbital elements of a satellite determine its flight track. The ephemeris of a satellite can be calculated from its orbital elements. The ephemeris can be regarded as a function of time, which represents the position of the satellite at any moment and the magnitude and direction of the instantaneous velocity. The ephemeris is usually given in the form of a table, and the data in the table are calculated from the orbital elements combined with Kepler's laws. This process is a mature technology in the aerospace field and is not the focus of this book, so the satellite's orbital elements are given in the experiment, and the ephemeris information of the satellite can be obtained according to the orbital elements. The satellite orbital elements chosen for the experiment are the Kepler orbital parameters, which detailed in Table 4.2.

Table 4.2: Orbital elements of the satellite KD-1.

Parameter	Value	Parameter	Value
semi-major axis a (km)	7200	Ascending intersection equidistant Ω (°)	175
Eccentricity e (°)	0	Perigee angle ω (°)	0
Orbital inclination i (°)	96.6	Flat perigee angle M (°)	0

The satellite in the simulation experiment belongs to the low-Earth-orbit circular satellites, and its flight time of one cycle around the Earth is about 90 minutes. The satellite ephemeris obtained from the satellite's orbital elements is the basis of many subsequent calculation processes, such as the calculation of the time windows and pointing angles of tasks, and the calculation of the attitude transition time between two tasks.

2) *Attitude maneuvering of the satellite*

The satellites designed in this experiment are agile imaging satellites, i. e., the imaging payload can adjust the pointing angle within a certain range according to the geographic coordinates of the task and the position of the satellite, in order to realize more flexible scheduling of the imaging tasks. The attitude maneuver capability of the satellite is particularly important for the visibility calculation of the imaging satellite, which is the basis for calculating the visible time windows of tasks. The satellite's attitude maneuver capability is summarized in Table 4.3.

Table 4.3: Attributes about the satellite attitude maneuvering capability.

Variable	Maximum value (°)	Minimum value (°)
Roll angle	45	−45
Pitch angle	45	−45
Yaw angle	0	0

The roll angle is the acute angle of rotating along the forward direction of the satellite; the pitch angle is the acute angle of rotating along the axis perpendicular to the forward direction of the satellite and parallel to the ground, with the satellite as the vertex. Therefore, the satellite's roll angle is positively correlated to the offset of the task position from the subsatellite track, and the pitch angle is positively correlated to the offset of the task from the subsatellite point in the forward direction of the satellite. Because the satellite designed in this experiment does not have the ability to actively push sweep and imaging satellite in motion, the design yaw angle is zero.

   3) *Other capabilities and parameters*

   Based on the design of the input resource parameters in Subsection 3.3.1, the set **RS** is used in the experiment to define the imaging resource capacity parameters for the satellite scheduling problem. Combined with the description of the model (3.28), it can be sorted out that the resource capacity to be considered in this scheduling problem is mainly the cumulative imaging duration not exceeding 3600 seconds. This constraint is prevalent in engineering practice, and its data is generally calculated by combining the satellite's solid-state memory constraints, power constraints, and other conditions. In addition, the specific parameters of imaging clarity, attitude transition time, and related calculations have been given and analyzed in detail in Sections 4.2.2 and 4.2.3 of this chapter, so we will not repeat them here.

(2) Task parameter

The task information can be represented by the set **TS**, which contains the visible time window information of the task [ws, we], imaging duration d and profit p. In order to try to keep the simulation experiment consistent with the situation in the real problem, the input information is the set of imaging requirements. The simulation experiment starts from obtaining the imaging requirements and a series of computations based on the task pretreatment process to get the imaging task, which is used as one of the input information for the imaging satellite scheduling problem.

   Two strategies are used to generate the geographic locations of the tasks: "Chinese region," which means that all the tasks are randomly distributed in China, and "Global region," which means that all the tasks are widely distributed worldwide. Specifically, tasks in China are randomly generated in a rectangular area with a longitude between 73°E and 133°E, and a latitude between 3°N and 53°N, which obey the uniform distribution, i. e., Lon = $U(73, 133)$E, Lat = $U(3, 53)$N. The accuracy of the task in the global distribution encompasses the full domain longitude, i. e., from 180°W to 180°E. Due to the coverage of the satellite orbits, the distribution of the latitude ranges from 65°S to 65°N. The distribution obeys a uniform distribution, i. e., Lon = $U(-180, 180)$, Lat = $U(-65, 65)$. The generation of task geolocations is summarized in Table 4.4.

Table 4.4: Generation rules for two types of tasks.

Category	Latitude range	Longitude range	Probability distribution
Chinese region	[3°N, 53°N]	[73°E, 133°E]	Evenly distributed
Global region	[65°S, 65°N]	[180°W, 180°E]	Evenly distributed

In order to test the performance of HADRT and DPTS under different task sizes and distributions, a total of 20 sets of task sets are designed for the experiment. The task size and distribution of each group are shown in Table 4.5.

Table 4.5: Task size and distribution of experimental groups.

Experimental group	Task size	Distribution	Experimental group	Task size	Distribution
1	100	Chinese	11	200	Global
2	200	Chinese	12	400	Global
3	300	Chinese	13	600	Global
4	400	Chinese	14	800	Global
5	500	Chinese	15	1000	Global
6	600	Chinese	16	1200	Global
7	700	Chinese	17	1400	Global
8	800	Chinese	18	1600	Global
9	900	Chinese	19	1800	Global
10	1000	Chinese	20	2000	Global

When the task is located in the "Chinese region," all the tasks in the experimental group are geographically concentrated, and their visible time windows are clustered, which can simulate the scenario of geographically concentrated tasks; when the task is located in the "global region," the scenario of uniform distribution of visible time windows of the tasks can be simulated. Different task sizes represent the overall density of tasks, so the experiment can test the performance of the algorithm under different task distribution and different task sizes.

Therefore, the raw input form in the experiment is shown in Table 4.6. Among them, information such as profit, imaging sharpness requirement, and imaging duration are generally provided by the user in conjunction with the purpose of using the satellite image. In the simulation experiment, these parameters are generated randomly, and the rules for generating random data are shown in Table 4.7.

Table 4.6: Imaging requirements (example).

Requirement No.	Profit	Imaging clarity requirement	Imaging duration	Geographic coordinates	Type of requirement
1	7	8	16	25.95°N, 120.39°E	point target
2	4	5	25	41.20°N, 117.31°E	point target
3	10	5	15	46.65°N, 110.44°E	point target
4	10	7	21	45.27°N, 79.91°E	point target
5	3	9	22	4.36°N, 117.20°E	point target
...	...	...	...	...	...

Table 4.7: Requirement-related parameter generation rules.

Attributes	Minimum value	Maximum value	Distribution	Variable type
Profit	1	10	Uniformly	Integer
Imaging quality requirements	5	9	Uniformly	Integer
Imaging Duration	15	29	Uniformly	Integer

Combining Table 4.2, Table 4.3, and Table 4.6, which computes the visible time window for each imaging task and the pointing angle of each task at any point in time within the time window, is shown in Table 4.8 and Table 4.9.

Table 4.8: The visible time window of tasks (example).

Task ID	Time window ID	Start time	End time	Length
1	1	2013-04-20 10:46:28	2013-04-20 10:46:42	14
2	1	2013-04-20 22:00:45	2013-04-20 22:06:47	362
3	1	2013-04-20 10:43:55	2013-04-20 10:49:04	309
...	...	...	...	...

Table 4.9: Satellite pointing angle to the task at each moment point in the time window (example).

Task ID	Time window ID	Time	Roll angle	Pitch angle
1	1	2013-04-20 10:46:28	44.324	44.964
1	1	2013-04-20 10:46:28	44.375	44.859
...	...	...	...	...
1	2	2013-04-20 22:00:45	−32.156	44.877
...	...	...	...	...

It is worth noting that in this problem, only the efficiency of each algorithm in the task scheduling problem is discussed. Therefore, in Table 4.8, there is only one time window corresponding to each task, i. e., each task has only one imaging opportunity to simplify the task assignment process.

At this point, the input parameters necessary for the imaging satellite scheduling problem and the way they are generated have all been presented. These parameters, combined with the constraints and objective function of the problem, can be used to obtain a solution scheme through the corresponding algorithm.

4.5.2 Experimental results and analysis

This experiment tests the efficiency of HADRT and DPTS when solving the imaging satellite task scheduling problem—the learning-based ant colony optimization (LACO) [122], adaptive large neighborhood search (ALNS) [14], and another algorithm, heuristic algorithm based on time-window starting time of task visible time window (HATW), as comparison algorithms. The performance of the proposed algorithms in each scenario is comparatively analyzed. HATW is also one of the commonly used heuristic algorithms in the task scheduling process of imaging satellites in practical engineering, and its basic principle is that: the task scheduling is sorted in accordance with the start time of the visible time window of the task, and the start time of the task is determined one by one in accordance with the tight scheduling strategy, with checking the constraints.

(1) Task completion rate and profit rate

The experiment begins with statistics and analysis of the task completion rate and task profit rate of each algorithm in different scenarios. Task completion rate and task profit rate are statistical quantities obtained from further processing of a task scheduling scheme, and both are common measures of the quality of a scheduling scheme. Task completion rate is the ratio of the number of successfully scheduled tasks to the total number of tasks, which is calculated as shown in equation (4.36); task profit rate is the ratio of the sum of the profits of the successfully scheduled tasks to the sum of the profits of all the input tasks, which is calculated as shown in equation (4.37):

$$\text{Task completion rate} = \frac{\text{Number of tasks in the scheme}}{\text{Total number of the task set}} \times 100\,\% \tag{4.36}$$

$$\text{Task profit rate} = \frac{\text{Total profit in the scheme}}{\text{Total profit of all candidate tasks}} \times 100\,\% \tag{4.37}$$

The task completion rate and task profit statistics of each algorithm in the experimental scenarios are shown in Table 4.10. The table does not record the results corresponding to experiments where the program running time exceeds 3600 s, because the satellite operation control department wants the algorithm to get a better task scheduling scheme in the shortest possible time during the task scheduling process. Analyzing the data in the table, it is not difficult to find:

Table 4.10: Task completion rate and task profit rate statistics.

Task set	Task completion rate (%)					Task profit rate (%)				
	HADRT	DPTS	HATW	LACO	ALNS	HADRT	DPTS	HATW	LACO	ALNS
Chinese_100	100	100	100	100	100	100	100	100	100	100
Chinese_200	100	100	100	99.5	100	100	100	100	99.6	100
Chinese_300	100	100	100	96	97.6	100	100	100	95.5	95.8
Chinese_400	100	100	100	94.5	91.6	100	100	100	93.7	92.6
Chinese_500	98.8	100	98.9	85.4	84.3	98.7	100	99.1	86.1	89.2
Chinese_600	96.7	100	95.5	76.3	75.9	96.9	100	95.7	80.8	85.9
Chinese_700	93.3	99.2	90.5	–	69.2	92.8	99.3	90.3	–	76.7
Chinese_800	87.7	92.3	85	–	65.5	86.6	94.5	83.9	–	75.1
Chinese_900	82.3	88.6	79.2	–	59.2	81.6	90.2	78.4	–	68.7
Chinese_1000	79.3	85.3	74.2	–	55.4	79.8	88.7	73.2	–	62.8
Global_200	100	100	100	100	100	100	100	100	100	100
Global_400	100	100	100	100	100	100	100	100	100	100
Global_600	100	100	100	100	100	100	100	100	100	100
Global_800	100	100	100	100	100	100	100	100	100	100
Global_1000	100	100	100	–	100	100	100	100	–	100
Global_1200	100	100	100	–	100	100	100	100	–	100
Global_1400	100	100	100	–	100	100	100	100	–	100
Global_1600	100	100	100	–	100	100	100	100	–	100
Global_1800	100	100	100	–	100	100	100	100	–	100
Global_2000	100	100	100	–	100	100	100	100	–	100

① In all the scenarios, DPTS can achieve a task completion rate and task profit rate that is not lower than that of other algorithms. In the scenario "Chinese region_1000," which has the most task conflicts, DPTS achieves a task completion rate higher 30 % and a task profit higher 25 % than ALNS;

② Except for the scenario "Chinese region_500," HADRT is able to stably outperform the other three comparative algorithms, i. e., HATW, LACO, and ALNS, in all the simulation scenarios. In the scenario "Chinese region_500," HATW's task completion rate is almost equal to that of HADRT. Combined with the conclusions in (1), it can be shown that the two deterministic algorithms HADRT and DPTS designed in this book can obtain high accuracy in solving the imaging satellite task scheduling problem;

③ HATW can also obtain satisfactory solutions in most scenarios, but the gap between the task completion rates and task profit rates of HATW and HADRT and DPTS gradually widens with the increase of the task conflict degree in the scenarios. It indicates that the solution of HATW in complex scenes is worse than the two deterministic algorithms proposed in this book;

④ LACO can only obtain the final solution in scenarios with small task sizes, and cannot obtain the final solution in 3600 s in scenarios with large task sizes ("Chinese region_700" to "Chinese region_1000," "Global region_500" to "Global region_1000").

In conclusion, constructive heuristics usually get better solutions in specific problems, and DPTS and HADRT designed in this book are better than HATW, a commonly used heuristic algorithm in engineering; under the condition of limited number of iterations, metaheuristic algorithms such as LACO and ALNS give poor results, and are also prone to fall into the locally optimal solutions.

(2) Comparison of program running time

The experiment also compares the program runtime of each method. The program runtime represents the time efficiency of the program. Due to the time efficiency of the algorithms in real application scenarios, realizing the fast response of the algorithms plays a crucial role in improving the performance of the system. The maximum number of iterations for the metaheuristic algorithms LACO and ALNS is set to 5000, and when the algorithms determine that the results converge, the algorithm also ends the computation. Table 4.11 gives the statistics of the average program running time.

Table 4.11: Program runtime statistics.

Task set	Program run time (s)				
	HADRT	**DPTS**	**HATW**	**LACO**	**ALNS**
Chinese_100	1.19	10.46	**1.09**	13.2	1.8
Chinese_200	2.7	22.60	**2.51**	93.64	37.175
Chinese_300	4.54	39.64	**4.26**	224.21	339.64
Chinese_400	6.79	59.89	**6.33**	666.64	464.35
Chinese_500	9.86	103.97	**8.93**	1402.75	448.6
Chinese_600	12.39	161.36	**11.36**	3091.65	590.63
Chinese_700	15.93	227.95	**14.86**	>3600	692.62
Chinese_800	18.37	300.54	**17.09**	>3600	856.33
Chinese_900	23.11	374.22	**21.66**	>3600	1053.81
Chinese_1000	25.79	454.91	**24.4**	>3600	1186.15
Global_200	2.54	21.88	2.37	1.536	**0.38**
Global_400	6.2	47.55	5.8	94.28	**0.93**
Global_600	11.53	81.60	10.75	1639.02	**1.72**
Global_800	17.54	130.72	16.23	3257.41	**2.67**
Global_1000	25.38	227.71	23.64	>3600	**4.01**
Global_1200	33.74	349.33	32	>3600	**5.49**
Global_1400	43.58	501.31	41.6	>3600	**11.88**
Global_1600	55.48	650.99	**53.53**	>3600	364.54
Global_1800	68.89	759.83	**66.5**	>3600	750.61
Global_2000	84.36	993.67	**81.16**	>3600	1208.66

As can be seen from Table 4.11:

① The program runtimes of HADRT and HATW are close to each other and are maintaining a low level in all scenarios. Especially in the scenarios of "Global region_1600," "Global region_1800," "Global region_2000" and all "Chinese region," the running time of these two algorithms is close to the same level. Thus, these two algorithms outperform all other methods in terms of time efficiency.

② In all the scenarios, the program computation time of HADRT is slightly higher than that of HATW. This is caused by the fact that the scheduling success rate of HATW is lower than that of HADRT. After a task has been successfully scheduled, there is a small amount of computational process used to update the states of the task and scheme.

③ Although the running time of DPTS is much higher than that of the two heuristic algorithms, its computation time does not explode with the increase in task size, and the program running time is within an acceptable range when facing a scheduling scenario with no more than 2000 tasks, and is more stable in terms of time than that of the ALNS algorithm.

④ The LACO algorithm takes the longest time among all the scenarios. The running time of the algorithmic program of LACO shows exponential growth with the growth of the task size. Therefore, the LACO algorithm is not suitable for solving large-scale scenarios.

⑤ Algorithm ALNS overall requires more computing time than algorithms DPTS and HADRT. It is worth noting that in the "Global region_200" to "Global region_1400" scenarios, the algorithm ALNS has less algorithmic runtime than the other algorithms. This means that ALNS can converge quickly when dealing with scheduling problems with low task densities, whereas as the size of the problem solution increases and the difficulty of the solution increases, the running time of ALNS is often difficult to guarantee and rises steeply as the solution size increases. HADRT in all task sets is within acceptable time limits.

In conclusion, the computation time of DPTS and the two heuristics is manageable and acceptable in all scenarios, especially in scenarios with large task sets, where HADRT and HATW are more advantageous.

Summarizing the above experimental results, the following conclusions can be obtained: the DPTS is able to obtain a high-quality solution scheme in a controlled time, and HADRT is able to obtain a satisfactory solution in a shorter computing time. Therefore, in the scenarios with a significant task conflicts, the DPTS can be utilized to effectively improve the quality of the solution, while in the scenarios with a small degree of conflict, the choice of HADRT algorithm can guarantee the quality of the solution under the premise of achieving high computational efficiency.

(3) Correlation analysis

In order to corroborate the obtained conclusions, the experiment further analyzes the correlation between the selection of different algorithms, the program computing time, task completion rate, and task profitability, as shown in Table 4.12. The correlation analysis can indicate the degree of association between the attributes, and if the absolute value of the correlation between the algorithm and the evaluation indexes is closer to 1, then it can indicate that the algorithm is effective in improving the corresponding indexes.

Table 4.12: Correlation analysis.

	Algorithm	Program computation time	Task completion rate	Task profit rate
Algorithm	1	−0.9228	−0.9927	−0.9886
Program computation time	−0.9228	1	0.05067	0.07381
Task completion rate	−0.9927	0.05067	1	0.99236
Task profit rate	−0.9886	0.07381	0.99236	1

The data in the table shows that there is a high correlation between the algorithms and the program computing time, task completion rate, and task profit rate, which indicates that different algorithms have differences between the program computing time, task completion rate, and task profit rate, and have a strong regularity on different data sets, which have been analyzed through the experimental results in this chapter. Task profitability and task completion rate have a high correlation, which is determined by the properties of these two metrics. The correlation between the program computing time and the task completion rate and task yield is not strong, which shows that the length of the algorithm computing time has no direct association with the scheduling scheme profit.

4.6 Summary

Based on the typical agile satellite task scheduling problem, the mathematical planning model and deterministic algorithms for task scheduling are studied in this chapter. On the basis of analyzing the mathematical planning model for task scheduling, the simplification and elimination of two classes of typical constraints in the imaging satellite task planning problem (imaging quality constraints and attitude transition time constraints) are realized, and the simplification strategy for multiple classes of constraints are summarized to reduce the number of constraint entries; through the analysis of the summarized four classes of constraints and processing, a constraint checking algorithm based on the timeline advancement mechanism is proposed to realize an efficient constraint checking process, which is the basis of the algorithmic research in this chapter.

Compared with stochastic optimization algorithms, deterministic algorithms are computationally efficient and the quality of solutions can be guaranteed, so they have important practical significance in real engineering problems. Based on the analysis and processing of the imaging satellite scheduling model, this chapter designs two deterministic algorithms to solve the task scheduling problem: the heuristic algorithm based on the density of residual tasks (HADRT) and the dynamic programming algorithm based on task sequencing (DPTS). The optimality of the two algorithms under specific conditions can be guaranteed theoretically, and the time and space complexity of the two deterministic algorithms are analyzed to illustrate the feasibility and effectiveness of the algorithms.

In order to verify the effectiveness of HADRT and DPTS in the imaging satellite task scheduling problem, this chapter also designed 20 sets of task scheduling scenarios to carry out simulation experiments. The task completion rate, task profit rate, and program running time of HADRT, DPTS, and the other three algorithms are compared in each scheduling scenario, and the validity of the conclusions is verified by the correlation analysis of each index. The experimental data show that among all tested algorithms, DPTS has the highest solution accuracy, especially in scheduling scenarios with high task conflict, but DPTS is no advantage in terms of computing time when dealing with large-scale tasks; HADRT can obtain satisfactory task completion rates and task profit rate at a small cost of computing time in most cases. The proposed DPTS and HADRT in this chapter show higher solution accuracy and solution stability than the other three algorithms in all scenarios, thus illustrating the advancement of the two proposed algorithms.

In the actual application process, imaging satellite managers can first make a preliminary judgment of the satellite task scheduling scenarios and select the corresponding algorithms based on the characteristics of the scenarios: in the scenarios where the tasks are more centralized and the higher degree of task conflicts, priority is given to the use of DPTS to ensure the algorithm's solution accuracy, and in the scenarios where the tasks are more dispersed and the task scale is larger, HADRT can be selected to obtain a high-quality solution while satisfy the requirements of the management and control side for the high timeliness of the computing process.

5 Research on task assignment problems of imaging satellites based on reinforcement learning

This chapter focuses on the properties and methods of the imaging satellite task assignment process. The difficulty lies in applying the characteristics of the problem to design the finite Markov decision (MDP) model and reinforcement learning algorithm realistically to improve the convergence and generalization of the algorithm in the training process, as well as the efficiency and accuracy in the testing process. First, the elements of the MDP model oriented to the task assignment process are designed in detail to fit the problem. Then an improved deep-Q learning algorithm (DQN) is proposed, which integrates knowledge from the imaging satellite task planning field into the training framework and pruning strategy to improve the algorithm's training efficiency and solution accuracy. Finally, based on many experiments, the generalization, convergence, and computational efficiency of the algorithm are analyzed, while the feasibility and effectiveness of the algorithm are verified. The progressiveness of learning-based bilevel methods in solving imaging satellite task planning problems has been revealed.

5.1 Literature review

5.1.1 Task assignment models and algorithms

In the field of imaging satellite task planning, task assignment is usually combined with task scheduling algorithms to work together, and the imaging satellite task assignment process will not be discussed separately from the task scheduling process. The task scheduling algorithms discussed in the previous chapter treat the problem as a whole for processing and optimization, however, as the scale and difficulty of the imaging satellite task planning problem increases, especially the number of satellites, the "combinatorial explosion" often occurs. At this time, "divide and conquer" is an effective means to reduce the satellite task planning problem's complexity and improve the solution's efficiency. There is no shortage of real-world examples of complex planning and scheduling problems described as bilevel optimization problems: the personnel scheduling problem [136, 137], the transportation domain [138], and the multimachine scheduling problem in industry [112, 139], among others. The basic idea of solving these problems is first to determine the task assignment scheme through intelligent computational methods to reduce the problem's difficulty, followed by further utilizing task scheduling algorithms to form the final task planning scheme.

These examples are sufficient to demonstrate that bilevel optimization models can effectively reduce the solution complexity in real complex combinatorial optimization problems. In the field of imaging satellite task planning, the bilevel optimization model is often used in the multisatellite cooperative task planning problem [140, 141]. However,

https://doi.org/10.1515/9783111585109-005

these works are frequently application specific, which is not conducive to generalized applications or scientific research. For the task assignment process in the imaging satellite task planning problem, which is the upper-level optimization process of the bilevel optimization model of the imaging satellite task planning problem, there are two main implementations [142]: the task assignment algorithm based on the metaheuristic algorithm and the rule-based task assignment algorithm. The solution framework of the task assignment algorithm based on the metaheuristic algorithm is shown in Figure 5.1.

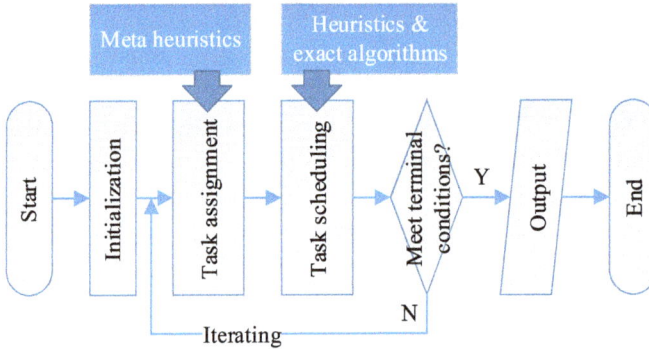

Figure 5.1: Framework for solving bilevel optimization models based on metaheuristics.

Chong Wang [143] used an improved fast simulated annealing algorithm to solve the satellite collaborative optimization decision-making problem and improved the solution quality and computational efficiency of the algorithm; Guoliang Li [142] proposed a multisatellite collaborative method based on the contractual network protocol, which through the bidding mechanism, will be used for every task to realize the task assignment process based on the scoring results of different satellites in the bidding process; Zailiang Yu [144] designed a multicenter collaborative task assignment model and proposed a decommissionable contract network task assignment method based on the integrity mechanism; Yongxian Chen [92] combined the genetic algorithm and simulated annealing algorithm to achieve robust planning for collaborative task planning of imaging satellites. Jiting Li [24] added a heuristic rule based on priority ranking to the traditional multiobjective genetic algorithm to solve the problem of autonomous collaborative task planning for multiple stars in high and low orbits, which bridges the gap in the domestic problem of autonomous collaborative task planning in high and low orbits.

The second implementation is a heuristic rule-based task assignment process, as shown in Figure 5.2. Yue Miao [145] designed a multisatellite hierarchical meritocracy algorithm to implement the task assignment process, which is essentially a constructive heuristic algorithm, i. e., it introduces multiple ranking rules for the objectives and selects the scheme with an enormous profit among these rules. By introducing the adaptive assignment algorithm, Lei He [100] can decompose the original complex multisatel-

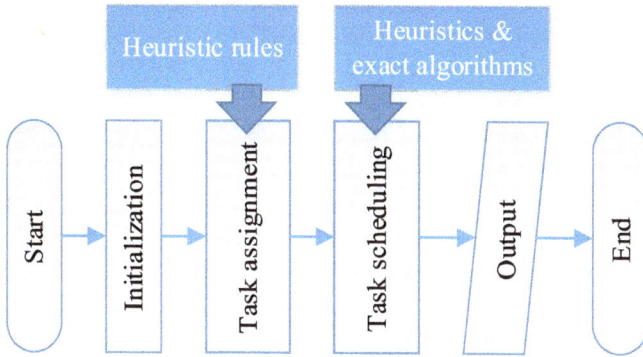

Figure 5.2: Framework for task assignment based on constructive heuristics.

lite task planning problem into a collaborative task assignment problem and multiple single-satellite planning subproblems, improving problem-solving efficiency. Among them, the adaptive assignment operator is essentially the construction of heuristic rules. Romain [146] proposed a task planning problem when a multiagile satellite system receives an urgent task in the middle of its execution, taking into account the balance between the computation time, the stability of the solution, and the optimality of the planning scheme.

From the above work, it can be concluded that more scholars prefer metaheuristic algorithms for solving task assignment problems because they can optimize the quality of the solution through continuous iterations. However, the limitations of both ideas in solving the imaging satellite task assignment problem are also obvious [147]: although constructive heuristic algorithms are fast in solving and easy to obtain satisfactory or even near-optimal scheduling solutions, their solving effect is limited by the quality of the heuristic rules constructed; although metaheuristic algorithms usually show a better-solving effect on large-scale problems, they cannot guarantee to find an optimal solution to the problem [148]. Therefore, designing a general and efficient bilevel optimization framework and designing algorithms under this framework to ensure the efficiency and accuracy of the solution process is of great significance in better studying the complex, large-scale imaging satellite task planning problem.

5.1.2 Applications of machine learning to combinatorial optimization

As we all know, machine learning is a methodology that utilizes the relationship between data, the laws embedded in data, and models to enhance the application effect of complex systems. It can summarize the knowledge and rules helpful in solving the target problem through statistical learning of large amounts of data and achieve fast and accurate decision-making [149, 150, 151] with the help of this knowledge or rules in the problem-solving process. Supervised learning, unsupervised learning, and reinforcement learning are the three significant paradigms of machine learning, of which unsupervised learning is widely used in clustering, pattern recognition, and feature extraction and is good at making decisions in scenarios where there is not enough a priori knowledge, it is good at clustering and labeling a large amount of data without given labels to find appropriate clustering features and labeling criteria, and few works apply unsupervised learning directly to guide decision-making. Supervised learning can generate empirical formulas to support decision-making, e. g., AlphaGo, by collecting human Go game records as labeled data and summarizing the probability of choosing different drop points to win in various states from them [152]. Xiaogeng Chu [103] utilizes supervised learning to train satellite target decision-making models in task planning for imaging satellites. By preparing training samples in advance and training to obtain a classification model, imaging satellites can decide whether to accept a specific task based on the model, thus reducing the calculation of the scheduling process. The disadvantage of this approach is that it requires a large amount of labeled data, and preparing this data is usually laborious and tedious. Moreover, the results of supervised learning are closely related to the characteristics of the data set used [153, 154], and how well this knowledge fits the attributes in a real-world problem depends on the objectivity of the collected data set.

Reinforcement learning is an important branch of machine learning methods that allow agents to interact with their environment and autonomously summarize empirical formulas for decision-making from them, without the need to prepare labeled data in advance, and has appeared in an increasing number of cases in the field of intelligent decision making, e. g., in the areas of Texas Hold'em [155], the game of Go [152, 153], ride-hailing order assignment [156, 157], and three-dimensional packing problems [158], all of which have seen world-beating successes of reinforcement learning in recent years.

Current mainstream imaging satellite task planning models share a common limitation: problem-solving efficiency is tightly coupled with the model's objective function, constraints, and even the range of variable values. When these conditions change, targeted processing and manipulation [159] of the model is usually required to achieve an efficient solution. This process is sometimes very complex and time-consuming. MDP models can effectively decouple the constraints from the decision-making process, and its core idea is to generate empirical formulas for decision-making through training, which is not concerned with the specific internal characteristics of the model [160]. Compared to mathematical models in stochastic process theory, such as Markov chains, Hidden Markov Models, and Semi-Markov Models, MDP emphasizes its Markovian nature, i. e., each state transition depends on only one previous state, and the empirical formulas are trained by continuous iteration through the agents and the environment during the solution process of the MDP model [161].

As the size of the problem increases, reinforcement learning also has its problem, such as low convergence efficiency. One idea is hierarchical reinforcement learning,

i. e., the problem is divided into several solution stages, and each stage designs the MDP model separately and adopts reinforcement learning algorithms to train the decision-making agents in the corresponding stages. The basic framework of hierarchical reinforcement learning is shown in Figure 5.3.

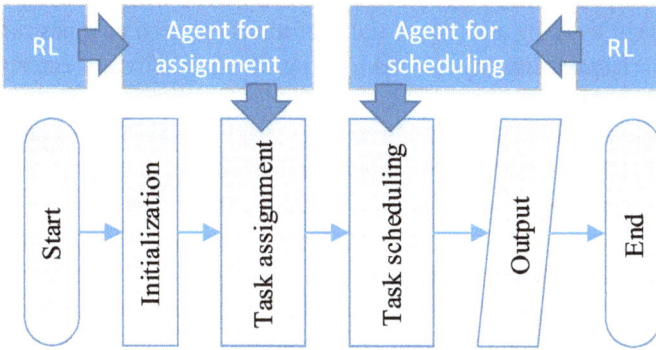

Figure 5.3: Framework for task assignment based on hierarchical reinforcement learning.

Describing a real-world problem as an MDP model is the first step in solving them by applying reinforcement learning. It is a skillful task, and the model's goodness profoundly affects the final solution. However, the lower-level task scheduling problem is well characterized to an combinatorial optimization problem. It can be solved by reasonably designing the algorithm to ensure the solution's quality and efficiency without the need to model it as an MDP model and use reinforcement learning to train its solving rules. Therefore, using a hybrid bilevel optimization model and integrating deterministic algorithms and reinforcement learning to solve the imaging satellite task planning problem can achieve an efficient and accurate solution to the problem.

The vigorous development of machine learning theory and applications has led to many breakthroughs in the modeling of planning and scheduling problems in MDP: the popularity of temporal difference algorithms in the field of planning and scheduling represented by Q-learning [162] and the successful application of "end-to-end" reinforcement learning algorithms represented by pointer-networks [102] in VRP problems, underwater automated piloting control [163], and Unmanned Aerial Vehicle control [164, 165], etc. Luca et al. [166] proposed the ant-Q algorithm for solving the Traveling Salesman Problem (TSP); Wei and Zhao [167] used Q-learning to train composite rules for selection (machine-job pairs) and applied it to the JSP problem; Khalil et al. [168] utilized GNN-based reinforcement learning algorithms for training a value function for each decision-making step in the TSP problem value function and implement the decision-making process for each step in the TSP problem based on this value function; Nazari et al. [39] solved the VRP problem using pointer networks; and Li [169] applied an improved actor-critic algorithm to train deep neural network models for solving multiobjective TSP problems. These works show that reinforcement learning has great potential for solving planning and scheduling problems. However, when applying reinforcement learning models and algorithms to complex problems, the convergence and generalization of the algorithms decrease dramatically within the same computational resources and computation time [170], which may be unacceptable in many practical applications. Therefore, decomposing the problem and solving the subproblems separately using reinforcement learning algorithms and operations research methods can reduce the solution search space and improve the training efficiency of reinforcement learning algorithms.

In addition, some studies have applied reinforcement learning to satellite task planning: Usaha and Barria [171] compared actor-critic algorithms with value-based algorithms in the resource allocation problem of the LEO satellite system. They designed an explicit function to describe the value function. Chong Wang [172] solved the multisatellite cooperative task planning problem based on the blackboard model and multiagent reinforcement learning. Among them, the agent assigns tasks, and the blackboard model reduces communication costs. Haijiao Wang [173, 174] considered the online scheduling problem of image satellites as a dynamic stochastic knapsack problem and then solved the problem by the asynchronous advantage actor-critic (A3C) algorithm. However, the satellite scheduling problem in this work only considers the most basic task uniqueness constraints and many complex constraints in real-world problems are simplified. The problems in these works can be characterized as MDPs of simple form and can be solved by designing novel reinforcement learning algorithms. Still, these works do not consider the specific constraints and conditions in complex real-world problems. Including these conditions can lead to low efficiency of the corresponding reinforcement learning algorithms if the MDP model is not further designed for this feature. Therefore, the MDP model and the reinforcement learning algorithm must be developed by combining the essential features and domain knowledge of the imaging satellite task planning problem to realize a more efficient reinforcement learning training and application process.

5.2 MDP model for task assignment problems

Based on the design in Chapter 3, this book models the task assignment problem as an MDP model. The MDP model excels at sequential decision-making problems. For the task assignment problem, the object of the decision is the task. Tasks are added to the set of tasks to be scheduled individually based on their state at the decision moment. This section starts from the interaction process between the environment and agents of the MDP model. Then the main components of the MDP model for solving the task assignment problem are detailed: action, state, reward, and value function.

5.2.1 Framework

The relationship between the parameters of the task planning scenario, the task assignment scheme, and the final profit can be obtained by trying different task assignment strategies in a large number of training scenarios and then scheduling tasks based on the task assignment strategy, thus realizing the training process of the empirical formula used to solve the upper-level task assignment problem. It is necessary to continuously interact with each other through the upper and lower levels of the solving process to realize this process. In the MDP model designed in this book, the relationship between the upper-level task assignment problem and the lower-level task scheduling problem can be summarized as follows:

①　The task assignment problem-solving module is based on the action selection policy in reinforcement learning to realize the problem's solution. The action selection policy is an essential part of the reinforcement learning algorithm, whose input is the state parameter s, and the output is the action a selected based on the policy in the current state.

②　The output actions of the task assignment problem are the input of the task scheduling problem. In this MDP model, the action selection strategy for assigning tasks generates one action a_i at a time, i. e., to decide on the matching between a task and resources. The task scheduling module receives the task assignment scheme as inputs to generate a task planning scheme. The task assignment scheme consists of the action sequence from the beginning of the scene to the current stage.

③　The task scheduling problem solving module updates the state parameters S_{l+1} based on the obtained task planning scheme, for task assignment in the next step.

In addition to the above two modules, the MDP model also includes some other necessary functional modules. The functional composition of the MDP model is shown in Figure 5.4.

The interaction of the modules with the task assignment and task scheduling process is summarized below:

①　The value function is the empirical formula that needs to be trained by reinforcement learning. Value function updating is the core issue of value-based reinforcement learning algorithms. The trained value function can be used as a task assignment policy to solve the task assignment problem.

②　The effectiveness evaluation algorithm module is based on the task planning scheme to calculate the profit F of the scheme. Based on the difference between the two returns before and after, $F_{i+1} - F_i$, the short-term reward R_i corresponding to an action in task assignment can be calculated.

③　Historical training data sets and task assignment schemes are recorded in data storage modules that store the intermediate training process. Based on the historical training data S_i, S_{i+1}, a_i, R_i, which is generated by the constant interaction between

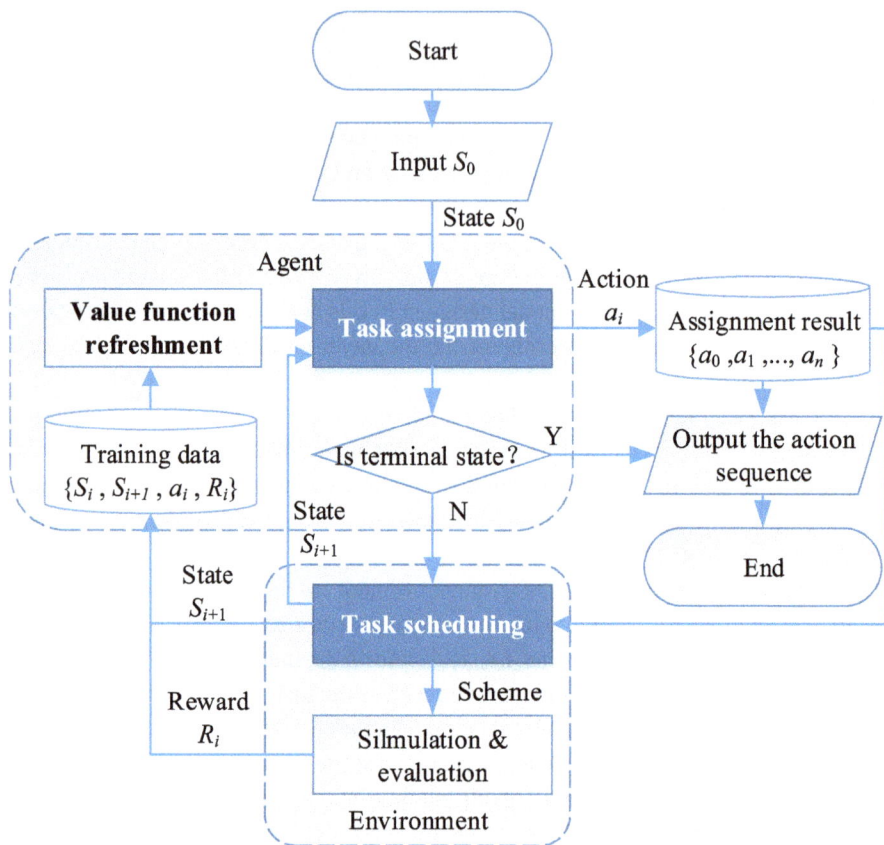

Figure 5.4: Functional composition of the MDP model.

the task assignment solving module and the task scheduling solving module, can be realized to update the action selection strategy continuously.

Therefore, the remainder of this section describes the reinforcement learning algorithm for solving the task assignment problem and devises the basic elements of the MDP model.

5.2.2 Action space

In this model, optional actions include "Select a task for the current resource (STR)" and "End adding tasks to the current resource (ETR)," which STR is the essential action in the action space. The set of all actions $\mathcal{A}_t$ before any time step t is shown in equation (5.1):

$$\mathcal{A}_t = \{a_0, a_1, \ldots, a_t\}, \quad a_i \in \mathbf{A}(S_i), \ i = 0, 1, \ldots, t \tag{5.1}$$

However, at each time step t, the feasibility of tasks changes dynamically with the change of states, i. e., not every task can be selected in all states. Therefore, the model designs some task screening conditions to ensure that every action in action chosen at decision time satisfies the basic constraints, reducing ineffective action attempts and improving training efficiency. Some of these conditions are related to the task attribute constraints of individual tasks (e. g., visibility constraints of the task in a particular track, etc.), and some are associated with the cumulative constraints or rolling constraints of the task (e. g., cumulative imaging duration constraints, cumulative number of maneuvers constraints, etc.). Common feasible action filters include and are not limited to:

① The task has not been selected at a previous time step, i. e., the task does not violate the uniqueness constraint if it is selected by this resource.

② The resource corresponding to the time step of the decision also has a visible time window for the task, i. e., the task does not violate the visibility constraint.

③ The solid-state memory consumption of the task to be joined should be less than the remaining solid-state memory of the current resource, i. e., it does not violate the satellite storage space constraint.

④ ...

During the study of the imaging satellite task planning problem, a resource defined in this book is an orbiting cycle of a satellite, so "end adding tasks to the current resource" indicates that no new tasks will be added to the current orbit, and the state jumps to selecting a task for the next orbiting cycle. If there is no feasible task at a time step t, then the only action available for the current state is the "end adding task for current resource." The design of this action is indispensable for the agent to jump out of the local optimum and find the global optimum. Figure 5.5 shows an example to illustrate this point.

In the instance shown in Figure 5.5, it is assumed that there are only five tasks in total and that the squares represented by the tasks contain two numbers, with the number in parentheses denoting the profit after executing this task, and the number outside of the parentheses denoting the task number. In Figure 5.5a, the sum of the profits of all the tasks in the first orbiting cycle before adding task 5 is 18 (assuming that the agent chooses tasks 1, 3, and 4), whereas adding task 5, the total profit increases to 20 (assuming that agent finally chooses tasks 1, 3, 4, and 5), but this will result in a profit of 0 for the second orbiting cycle because there are no more tasks to be executed. If the action "ETR" is available when there are still executable tasks, the possible result will become the situation shown in Figure 5.5b: when considering task 5, the agent can select "add task 5" or "ETR." If the agent can learn experience, adding task 5 can increase the total profit by 2, while selecting the action "ETR" can increase the total profit by 8, then the agent prefers "ETR" to "add task 5." Therefore, for each resource, if new executable tasks are added until there are no executable tasks, the total income of the current resource may be optimal, but the overall result is locally optimal. In the imaging satellite task planning

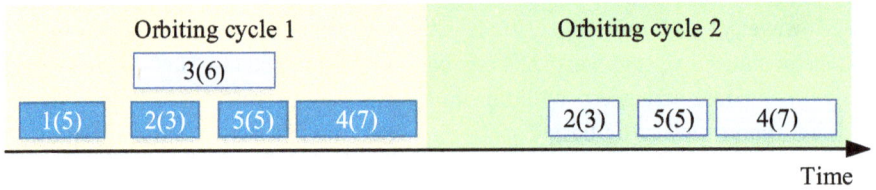

(a) without "ETR" in the action space

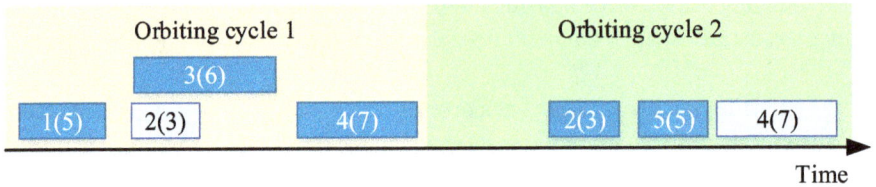

(b) with "ETR" in the action space

Figure 5.5: Example for explaining the difference between two settings in action space.

problem, a forward-looking agent is needed, which requires the action of "ETR" to be considered in the action space.

In summary, the action space in this MDP model is discrete, and its size is the number of tasks in the imaging satellite task planning model plus one:

$$|A| = N + 1. \tag{5.2}$$

5.2.3 State space

The state space S is the set of states describing the system. It can be represented as equation (5.3):

$$S = \{S_0, S_1, \ldots, S_i, \ldots\}. \tag{5.3}$$

S_i is the state at time step i, each of which is a collection of attributes describing the imaging satellite task planning problem at a particular stage during the solution. The choice of which attributes to describe the state also plays a key role in the training effectiveness of the model. Too little information in the state will lead to too much similarity of attributes describing different states, leading to underfitting of the training process; too many attributes describing the state will quickly lead to overfitting of the training process. Therefore, choosing attributes with high information density to portray the states in the MDP model in this paper can describe as much information as possible with the minimum number of state attributes. In the design of this model, some of the original attributes of the task requirements are selected to describe the state in combination with the attributes that are directly related to the decision-making process:

$$S_t = \{x_t^i = (g^i, p^i, v_t^i, l_t^i) \mid i = 0, 1, \ldots, n\} \tag{5.4}$$

g^i is the geolocation information of the task i, including the longitude and latitude of the imaging task. The geolocation information of the task can be formalized as follows:

$$g^i = (\text{lon}^i, \text{lat}^i) \tag{5.5}$$

Geolocation information was chosen because it is one of the task's fundamental attributes. After the task planning scenario has been defined, it can uniquely determine information such as the time window in which the task is visible on each satellite and the satellite's pointing angle toward the task at any given moment. Describing this relevant information in terms of geolocation maximizes the information density.

p^i is the profit obtained after executing the task i.

g^i and p^i are the original parameters submitted by the user when formulating the imaging requirement, and their values do not change during the task planning process.

v_t^i is the number of visible time windows remaining after the moment corresponding to the time step t for task i. This parameter must be calculated by the task pretreatment process. A task with no visible time window is called an invalid task, so an optional task in the action space of an arbitrary time step should have at least one visible time window remaining during its execution. The fact that a task can only be imaged within its visible time window is a hard constraint, so v_t^i is a necessary condition to be considered when making decisions. On the other hand, this attribute enables the agent to recognize how many observation opportunities different tasks have after one decision-making process, which facilitates the agent to make a more forward-looking decision scheme.

l_t^i is a token that records whether the task i can selected at time step i:

$$l_t^i = \begin{cases} 0, & \text{task } i \text{ at timestep } t \text{ is unselectable} \\ 1, & \text{task } i \text{ at timestep } t \text{ is selectable} \end{cases} \tag{5.6}$$

Several conditions under which a task i cannot be selected time step i have been analyzed during the design of the action space, where the visibility constraints and uniqueness constraints in combination with the design of the state-space variables can be described as formula (5.7) and formula (5.8):

$$v_t^i > 0 \tag{5.7}$$
$$\exists k \in \{1, 2, \ldots, t-1\}, \quad a_k = i \tag{5.8}$$

According to the above design, the state of any time step can described by S_i. In addition, the MDP model needs to define a rule to determine the Terminal state. The

Terminal state of the task assignment problem in the imaging satellite task planning problem can be determined by the formula (5.9), i. e., for all the tasks, there is no imaging opportunity after a certain time step t, and this state is the termination state:

$$v_t^i = 0, \quad \forall i \in \{1, 2, \ldots, n\}. \tag{5.9}$$

5.2.4 Short-term returns

The reward is a return value from the environment after an action is executed, used to measure the change in instantaneous profit after the action is completed. A positive short-term reward at any time does not necessarily mean the long-term value is positive. Still, by comprehensively training and fitting all the short-term reward data, the intrinsic connection between short-term reward and long-term value can be obtained, guiding the decision-making process. The set of short-term returns records the short-term returns from the environment at each time step:

$$\boldsymbol{R} = \{R_1, R_2, \ldots, R_t, \ldots\}. \tag{5.10}$$

Due to the specificity of the environment in this model, after each selection of an action, a series of complex calculations need to be performed in the environment by invoking the task scheduling algorithm to obtain the task planning scheme. So, the short-term reward in this model is not always equal to the profit value of the corresponding action but the difference between values of the objective function obtained before and after joining a task. Equation (5.11) gives its calculation:

$$R_t = F(\mathcal{A}_t, \mathbf{es}) - F(\mathcal{A}_{t-1}, \mathbf{es}), \tag{5.11}$$

where $F(\mathcal{A}_t, \mathbf{es})$ and $F(\mathcal{A}_{t-1}, \mathbf{es})$ are the values of the objective function at the time steps t and $t - 1$, respectively. Ideally, if there exists an algorithm to find an optimal solution of the Model (3.28), i. e., the short-term reward of the environment feedback is always the optimal value, then $F(\mathcal{A}_t, \mathbf{es})$ is the optimal total profit at time step i. However, due to the complexity of the problem, no algorithm can guarantee an optimal solution of the Model (3.28) in polynomial time, and thus, the short-term rewards of the environment feedback in the MDP model will not be the same with different algorithms for solving the lower-level task scheduling problem. The impact of this process on the overall problem solution will be discussed in detail in the experimental section of this chapter.

5.2.5 Value function

The value function is a core component of the reinforcement learning algorithm and an important criterion for selecting actions in the MDP model. It is a function of the combination of states and actions, and is used to evaluate the magnitude of the expected long-term profit of choosing different actions in a given state. Usually, according to Bellman's equation [151], the value function can be obtained by equation (5.12):

$$q(s, a) = E\left(R_{t+1} + \gamma \max_{a'} q(s_{t+1}, a') \mid S_t = s, A_t = a\right) \tag{5.12}$$

In the imaging satellite task assignment problem, since the state space is continuous and the action space is discrete, it is challenging to represent the value function based on spatially limited data structures, such as matrices, to accurately reflect the relationship between the state, the action, and the expected long-term value, thus making it challenging to support decision-making effectively. Artificial neural networks are superior in fitting and predicting complex functions compared to explicit functions [74, 175]. Therefore, in this scheme, a fully connected network is used to realize the description of the value function.

The value function based on a fully connected artificial neural network is shown in Figure 5.6. This fully-connected network is the most basic form of fully-connected networks and contains an input layer, a hidden layer, and an output layer. The inputs to the network are state parameters at time step t, and the output is a vector of expected long-term values corresponding to each action when the state is s. The network will eventually converge as it is trained by reinforcement learning algorithms. Ideally, in this problem, the network training process converges if the output vector is the maximum long-term value that can be obtained by taking different actions a any state s.

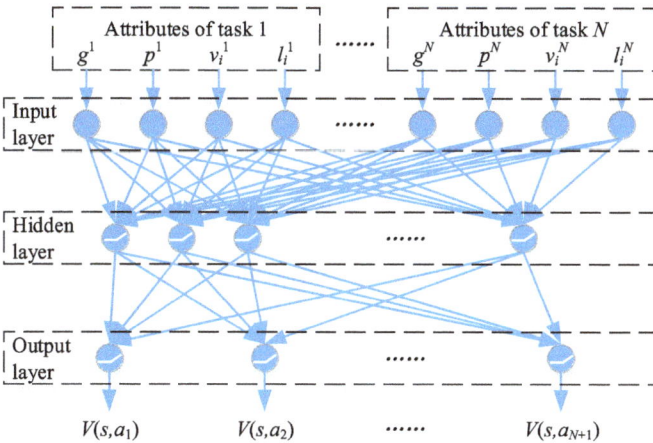

Figure 5.6: The value function based on fully-connected neural networks.

For the design of value functions based on fully connected networks, in addition to presenting the network's topology, the activation function, loss function, and optimizer in the network need to be designed.

(1) Activation function

The activation function is a crucial part of a neural network, which, together with other components, determines the relationship between the inputs and outputs of the neural network. With the activation function, the neural network is transformed from a linear model to a nonlinear one, giving it an advantage over linear functions in approximating more complex real-world situations. Most of the parameters in a neural network can be modified through training, but the activation function must be determined when building the neural network. This section organizes the mathematical expressions of the three classical activation functions, their characteristics, and their limitations.

1) *Sigmoid*

Sigmoid, also known as the logistic activation function, maps arbitrary real values in the interval of 0 to 1, and can also be used in the output layer of predictive probabilities. In this function, the smaller the input, the closer the function value is to 0, and vice versa. Its mathematical expression is

$$y = \sigma(x) = \frac{1}{1 + e^{-x}}. \tag{5.13}$$

The function image of Sigmoid is shown in Figure 5.7a. There are three main drawbacks of Sigmoid:

① Gradient vanishing: When the function value approaches 0 or 1, the gradient approaches 0. When a neural network applies Sigmoid for backpropagation, the gradient of neurons close to 0 or 1 approaches 0. These neurons are called saturated neurons. As a result, the weight update of these neurons is extremely slow and, at the same time, has a significant impact on the weight update process of the neurons connected to these neurons, which is known as gradient vanishing. So, imagine that if an extensive neural network with Sigmoid neurons, and many of them are saturated, the efficiency of the network in performing reverse propagation will be enormously reduced.

② The expected value of the function is not 0: The image of Sigmoid is a centrally symmetric graph with the center point (0, 0.5), i. e., the expected value of the function is not 0.

③ Computationally expensive: Exponential functions based on natural logarithms are computationally costly compared to other nonlinear activation functions.

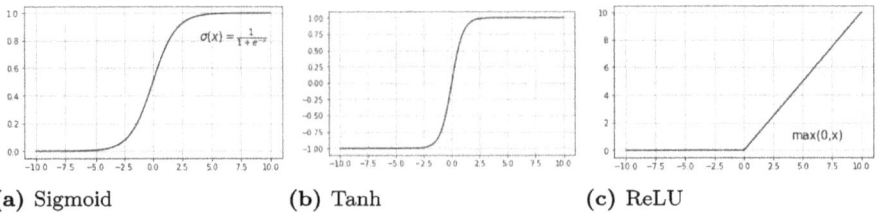

(a) Sigmoid (b) Tanh (c) ReLU

Figure 5.7: Function graphs of three different activation functions.

2) Tanh

Tanh activation function is also known as the hyperbolic tangent activation function. Similar to Sigmoid function, Tanh outputs true values of the function, but the Tanh function maps the values to the interval −1 to 1. Unlike Sigmoid, the expectation of the Tanh is zero because the interval is −1 to 1, and the function graph is symmetrical with the center point $(0, 0)$. In practice, Tanh function is used more frequently than Sigmoid. Tanh does not change the positive and negative properties of inputs. Its mathematical expression is

$$y = \tanh(x). \tag{5.14}$$

The function image of Tanh is shown in Figure 5.7b. Tanh also has the problem of gradient vanishing, but the gradient of Tanh vanishes when it is closer to saturation than Sigmoid.

3) ReLU

ReLU is also known as a rectified linear unit, which is a semirectified function. Its mathematical expression is

$$y = \max(0, x). \tag{5.15}$$

Therefore, the output of the ReLU function is 0 when the input $x < 0$, and the output is $x, x > 0$. This activation function allows the network to converge more quickly. It does not saturate in the positive real region (for $x > 0$), i. e., it fights against the gradient vanishing problem, so the neurons do not backpropagate all zeros in at least half of the region. Its function image is shown in Figure 5.7c.

ReLU is computationally efficient, but it also has some drawbacks:

① The image is not zero-centered. Similar to Sigmoid, the output of ReLU is not zero-centric.

② During forward passing, neurons become inactive and kill the gradient in backward passing (if $x < 0$). In this way, the weights cannot be updated, and the network cannot refresh. The gradient where $x = 0$ is not defined, so the gradient of this point is commonly replaced by the gradient of its proximity in practice.

(2) Loss function

The loss function describes a measure of the deviation between the training sample data and the output data of the neural network. It can be viewed as the objective function of the neural network training process. The training process continuously optimizes the parameters (weights) of the neural network to minimize the deviations between the sample data and the predicted data of the neural network. The loss function is the function used to describe these deviations. This section organizes several commonly used loss functions and their mathematical expressions:

① Minimum Squared Error (MSE)

$$\text{Loss} = \frac{1}{N} \sum_{k=1}^{N} (y_k - \widehat{y_k})^2 \tag{5.16}$$

② Minimum Absolute Error (MAE)

$$\text{Loss} = \frac{1}{N} \sum_{k=1}^{N} |y_k - \widehat{y_k}| \tag{5.17}$$

③ Mean Absolute Percentage Error (MAPE)

$$\text{Loss} = \frac{100}{N} \sum_{k=1}^{N} \left| \frac{y_k - \widehat{y_k}}{y_k} \right|^2 \tag{5.18}$$

④ Mean Squared Log Error (MSLE)

$$\text{Loss} = \frac{1}{N} \sum_{k=1}^{N} (\ln y_k - \ln \widehat{y_k})^2 \tag{5.19}$$

⑤ Hinge (HG)

$$\text{Loss} = \frac{1}{N} \sum_{k=1}^{N} \max(1 - y_k \widehat{y_k}, 0) \tag{5.20}$$

⑥ Squared Hinge (SH)

$$\text{Loss} = \frac{1}{N} \sum_{k=1}^{N} \max(1 - y_k \widehat{y_k}, 0)^2 \tag{5.21}$$

⑦ Binary Cross-entropy (BC)

$$\text{Loss} = - \sum_{k=1}^{N} (\widehat{y_k} \ln y_k + (1 - \widehat{y_k}) \ln(1 - y_k)) \tag{5.22}$$

⑧ Categorical Cross-entropy (CC)

$$\text{Loss} = - \sum_{k=1}^{N} (\widehat{y_{k1}} \ln y_{k1} + \widehat{y_{k2}} \ln y_{k2} + \widehat{y_{k3}} + \cdots + \widehat{y_{kn}} \ln y_{kn}) \tag{5.23}$$

⑨ Kullback–Leibler Divergence (KLD)

$$\text{Loss} = - \sum_{k=1}^{N} \widehat{y_k} \ln \frac{\widehat{y_k}}{y_k} \tag{5.24}$$

⑩ Mean Cosine Proximity (CP)

$$\text{Loss} = -\sum_{k=1}^{N}\left|\cos(\widehat{y_k}) - \cos(y_k)\right| \tag{5.25}$$

Each of the above loss functions has its characteristics. For example, MSE can usually converge faster than MAE, but MAE is more robust when encountering outliers; the hinge loss function performs better in convex optimization problems and is widely used in binary classification problems; the squared hinge loss function penalizes the outliers more severely than the hinge loss function, and so on. Most of the loss functions are chosen to be nonlinear, some of which are suitable for regression problems and some for classification problems, and this also needs further research. However, it is not a simple work to get the application law of each loss function in task assignment problems through mathematical derivation, and this chapter determines the loss functions suitable for task assignment problems utilizing simulation experiments.

The selection of the loss function and activation function are usually studied together because the backpropagation algorithm for chain derivation inevitably encounters the activation function, one of the most critical designs in forward propagation, which increases the diversity of the model and provides more nonlinear operations. Therefore, to find a neural network suitable for describing the relationship between decision and profit in the task assignment problem, the selection and matching of the activation function and loss function are of great significance in improving the stability, convergence speed, and solution accuracy of the algorithm in the solution process.

(3) Optimizer

Optimizers are used to find the function that minimizes the loss function. In this scheme, Stochastic Gradient Descent (SGD), Root Mean Square Propagation (RMSprop), and Adaptive Moment Estimation (Adam) algorithms are selected as three algorithms to investigate the effect of different optimizers on the value function learning process in the task assignment problem.

① The basic principle of the SGD algorithm is that there are three core elements of a gradient descent algorithm: the starting point, the descent direction, and the descent step size. The SDG algorithm is an improved version of the gradient descent method; unlike traditional gradient descent algorithms that need to calculate the gradient of all the starting points, updating the gradient of SGD each time is randomly selected a piece of data to be carried out. Hence, the most significant advantage of SGD is the efficiency and the ability to short time to realize a large number of iterations. For details of the implementation process, see [151].

② The basic principle of the RMSprop algorithm is that the RMSprop algorithm is based on a global learning rate that neutralizes the gradient variations in all the terms, such that the gradient of descent in all the terms converges to an average. This is accomplished by setting different coefficients for each operational term. The purpose of doing so is to increase the descent rate of points with gentle gradients while decreasing the rate of points with steeper gradients, increasing the overall training efficiency. See [161] for details of the implementation process.

③ The basic principle of Adam's algorithm is that Adam's algorithm can be regarded as an improved version of the RMSprop algorithm, which adaptively adjusts the learning rate of each parameter. The implementation process is detailed in [176].

Reinforcement learning produces a large amount of learnable data through continuous "Exploration" and "Exploitation," replacing the traditional process of labeling data in supervised learning. These data are used to train the value function. An experience replay mechanism is adopted to train the MDP model's value function [177]. The training process is detailed in Algorithm 5.1.

Algorithm 5.1 Training value function based on experience replay.

Input: The state at the current moment S_t, the state at the next moment S_{t+1}, the action at the current moment a_t and the corresponding profit r_t, the experience set ExpSet, the value network

Output: Updated value network Q

1: Add S_t, S_{t+1}, a_t, r_t into ExpSet;

2: Parameter initialization;

3: **for** $i = 1$: batch size **do**

4:     Load data in the experience set ExpSet.

5:     Stores the loaded data in the set InputSet;

6:     Computes $Q(S_t, a_t) = \max(\text{predict}(S_{t+1}))$;

7:     **if** S_{t+1} is the termination state **then**

8:         targets $\leftarrow r_t$;

9:     **else**

10:        Calculate targets from equation (5.12).

11:     **end if**

12:     Update the network parameters of Q based on the optimizer (SGD or RMSprop or Adam).

13: **end for**

Activation function, loss function, and optimizer also play an essential role in fitting the network. Due to the complexity of the imaging satellite task planning problem, these factors' impact on the training process is difficult to measure through theory. The experimental part of this chapter will expand in detail to discuss the effects of the loss function, activation function, and optimizer in the network on the solution results.

5.3 Improved deep Q-learning algorithm for solving the task assignment problem

Chapter 3 describes the reasons for choosing the "step-by-step" MDP model and its advantages. For the task assignment problem in imaging satellites task planning, although there is no random factor in the state transition, its initial state space is infinite, and the state sequence expanded for a specific scene is exponential with the increase of task size. Therefore, the model-based reinforcement learning algorithm is inefficient in dealing with this problem. In model-free reinforcement learning, temporal difference (TD) is one of the basic strategy. Q-learning and Sarsa are typical TD algorithms. The main difference between these two algorithms is that Sarsa is one of the on-policy algorithms. The strategy used to select actions during training is the same as that used to update the value function. In contrast, the Q-learning algorithm is a type of off-policy algorithm. Its strategy to select actions during training differs from its strategy to update the value function. Therefore, the learning process of Sarsa algorithm is smoother and more suitable for learning in a real online environment, such as the satellite autonomous task planning process [178]. But the Sarsa algorithm is more likely to fall into local optimization, so Q-learning is better for training the model in this book.

In this book, improved DQN is applied to solve the upper-level task assignment problem in the bilevel optimization model, in which deep Q-networks are used to represent the value function. However, standard DQN will still encounter new difficulties when solving the task assignment problem, which can be summarized as the following two aspects:

① Unlike the RL process for the game of Go, the initial states are different in this problem because user requests vary in various scenes. The standard DQN is trained intensely in one scene, which may lead to overfitting in the task assignment problems; that is, the value function obtained by training performs well in one scene but poorly in others. The primary problem that needs to be solved is making the algorithm fully trained in multiple scenes to avoid overfitting.
② In the MDP for task assignment, the feasibility of the task needs to be considered before "Exploration" or "Exploitation" in each time step. The algorithm design also focuses on how to fully use domain knowledge and problem settings to design pruning strategies for selecting actions.

The following content emphatically introduces the above two issues; thereby, the DQN for the task assignment problem is designed.

5.3.1 Solution framework

Ideally, the training process should traverse all initial states and all possible actions, then fit the desired value function based on all the records. However, there are infinite scenarios in the imaging satellite task planning problem. The state space corresponding to each scenario may have different characteristics, so the algorithm cannot realize the traversal of the state space, and the direct application of the classical DQN algorithm cannot effectively deal with the proposed MDP and task assignment problem. This book proposes a DQN training framework oriented to random initial states. Unlike the classical problems that DQN specializes in, the initial states of the imaging satellite task assignment problem are different in different scenarios, which poses new challenges to the training process. Embedding the process of randomly generating test scenarios into the training process of a deep Q-networks allows the network to learn through a wide range of scenarios, and thus generalize to unknown scenarios during the testing process.

Figure 5.8a shows the training process of the agent in reinforcement learning. The Q-network is updated once an action is taken and the corresponding reward is received. In each scenario, the agent runs through many scenarios and tries to select an action based on "Exploration" and "Exploitation" and then updates the Q-network through this process. The agent repeats this process in different scenarios and finally gets the value function for decision-making in any state.

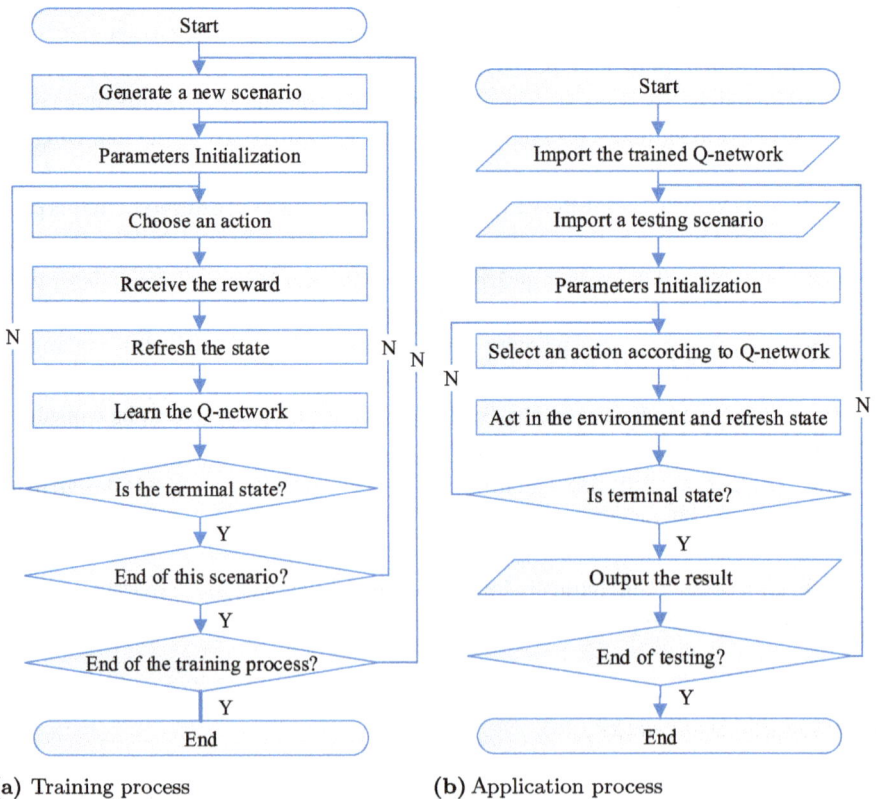

(a) Training process   (b) Application process

Figure 5.8: Improved Q-learning for the task assignment problem.

The main purpose of the testing process is to validate the effectiveness of the algorithm. The testing process is the same as applying the algorithm to real-world problems, with the difference that the inputs for the application are actual scenario data. In contrast, the test data for the validation process is randomly generated. The flowchart of the testing process is shown in Figure 5.8b. The difference between the training process and the testing process is summarized as follows:

① The testing process does not require multiple iterations of the same scenario.
② The Q-network will not be updated again during the testing process and, therefore, there is no longer a need to record historical data for each decision.
③ The strategy for task selection is also different: during testing, the agent always selects the action with the highest Q-value. In the training process, two strategies are used to choose actions: the "Exploitation" strategy selects the action with the highest expected value function, or the "Exploration" strategy randomly selects a feasible action.

5.3.2 Pruning strategy

In selecting actions, this book designs an efficient pruning strategy to prune nonexecutable tasks based on the domain knowledge and some constraints before each decision step to avoid the selection of ineffective tasks resulting in meaningless operations and to improve the overall computational efficiency of the algorithm.

The pruning strategy can be briefly described as follows: in the MDP model oriented to the task assignment problem, the overall action space is $N + 1$, and the time step is t. The actions can be filtered based on the following conditions to achieve the purpose of reducing the space of candidate actions:

① The task attribute constraints of the current time step t are unsatisfied. For example, the task does not have a visible time window at the current time step. The current time step does not satisfy the uniqueness constraint.
② When a task is added to the candidate task set at the current time step t, the cumulative constraints or rolling constraints in the task set are not satisfied.

More targeted pruning strategies can be designed to deal with the imaging satellite task planning problem. All in all, the introduced pruning strategy eliminates invalid actions, which can effectively reduce the decision space. The agent selects only available tasks each time to reduce the complexity of the constraint-checking process and improve the search efficiency. The following sections are oriented to the training and application phases, respectively, to illustrate in detail how the above pruning strategies work.

In the training phase, the pruning strategy supports the "Exploration" and "Exploitation" strategies in the DQN algorithm to improve efficiency. "Exploration" and "Exploitation" are important concepts in reinforcement learning. "Exploitation" tends to select the optimal action based on the currently obtained value function. In contrast, "Exploration" tends to choose an action that has not been attempted or has performed poorly in a similar state. Actions that have not been tried or have performed poorly in similar states. By randomly invoking one of these two processes, the value function is continuously modified until it converges.

The pseudocode for the training process with the above pruning strategy is shown in Algorithm 5.2.

Algorithm 5.2 DQN training process with pruning based on domain knowledge and constraints.

Input: The state at the moment of decision s, the set of actions A, the value network Q
Output: The selected action a

  1: Set the threshold ε for random selection of tasks;
  2: **for** a **in** A **do**
  3:   **if** action a is invalid **then**
  4:     Action a out of the list of available actions **AL**
  5:   **end if**
  6: **end for**
  7: index $\leftarrow$ randon();
  8: **if** index $< \varepsilon$ **then**
  9:   randomly selects a task from the list of all available tasks **AL** and outputs it.
 10: **else**
 11:   Calculate the result of the value network based on the current state s $Q(s)$;
 12:   **for** a **in AL do**
 13:     computes the Q values corresponding to all tasks in the set **AL**.
 14:   **end for**
 15:   selects the task corresponding to the largest value of Q in the set **AL** and outputs it.
 16: **end if**

When training with the above algorithm, a random number index is first randomly generated. Then the random number index is compared with a threshold: a feasible action is randomly selected if the index is smaller than the threshold. Otherwise, the feasible action with the highest Q-value in the Q-network obtained from the previous learning process is selected.

In the application phase of the algorithm, the algorithm no longer performs the "Exploration" process. The algorithm chooses the action with the largest corresponding Q-value for each selection action. The corresponding pseudocode is Algorithm 5.2 Stripped of the rest of lines 7 through 10. The pseudocode for the application phase will not be repeated due to content duplication.

5.3.3 Complexity analysis

(1) Time complexity
The computational process of the algorithm consists of two parts: the training phase and the testing phase. The time complexity of the training phase is discussed first.

From Figure 5.8a, it can be easily seen that the main process of the training phase includes action selection and training the network, where the time complexity of selecting an action is $O(N)$ and N is the number of input tasks. Assuming the size of the batch data is b, the time complexity of training the batch once TC_{NN} is

$$TC_{NN} = b(|S|n_{hid} + n_{hid}|A|) = O(N). \tag{5.26}$$

Discussing the performance of DQN algorithms for upper-level task assignment problems requires considering the impact of the performance of deterministic algorithms in the environment on overall performance. The computation in the environment consists of updating the state and feeding back rewards, which are mainly realized by the constructive heuristic algorithm based on the density of remaining tasks (HADRT) or the dynamic programming algorithm based on the task sequencing (DPTS), where the time complexity of HADRT is $O(N^2)$ and that of DPTS is $O(N^3)$. Assuming that the number of scenarios used for training is c, and the training is repeated for e generations in each scenario, TC_{act} is the time complexity of the main process of action selection, the time complexity of the TC_{NN} is the neural network update time complexity, TC_{env} is the time complexity of the relevant operations in the environment. Then the time complexity of the training process TC_{train} is

$$TC_{train} = ceN(TC_{act} + mTC_{NN} + TC_{env})$$

$$= \begin{cases} O(N^3), & \text{call HADRT for solving in environment} \\ O(N^4), & \text{call DPTS for solving in environment} \end{cases} \tag{5.27}$$

In which, TC_{act}, TC_{NN}, and TC_{env} refers to the time complexity of choosing actions, refreshing the networks, and calculating in the environment, respectively. Since the testing process only needs to call the result of the Q-network every time the action is selected, the time complexity of the testing process TC_{test} is

$$TC_{test} = N\left(TC_{act} + \frac{1}{b}TC_{NN} + TC_{env}\right)$$

$$= \begin{cases} O(N^3), & \text{call HADRT for solving in environment} \\ O(N^4), & \text{call DPTS for solving in environment} \end{cases} \tag{5.28}$$

Therefore, as the task size N increases, the training and testing time of the algorithm increases polynomially. Although the constants c, e, m, b, and n_{hid} also directly affect the computation time of the algorithm, this does not prevent the algorithm from being scalable in large-scale scenarios.

(2) Space complexity

The algorithm maintains an experience set and a network during the training process. In contrast, only the network needs to be maintained during testing. The role of the experience set is to store the training history data, which contains m records, each of which contains state s, state s', action a, and reward r information in time steps. The space complexity of the experience set SC_{exp} is computed as equation (5.29):

$$SC_{exp} = m|\{s, s', a, r\}| = 2m(5N + 1) = O(N) \tag{5.29}$$

According to the design in Section 5.2.5 of this chapter, the size of the state space and the action space is the number of nodes in the input and output layers of the network, respectively. Assuming that the number of nodes in the hidden layer is n_{hid}, and each node contains two parameters, the space complexity of the neural network SC_{NN} is computed as in equation (5.30):

$$SC_{NN} = 2(|S| + n_{hid} + |A|) = O(N). \tag{5.30}$$

Therefore, the space complexity of the training process SC_{train} is computed as in equation (5.31):

$$SC_{train} = SC_{exp} + SC_{NN} = O(N). \tag{5.31}$$

The space complexity of the testing process SC_{test} is computed as in equation (5.32):

$$SC_{test} = SC_{NN} = 2(n_{hid} + 6N + 1) = O(N) \tag{5.32}$$

5.4 Simulation experiment

The implementation of the simulation experiments in this chapter are divided into two parts, and conclusions of two aspects are discussed, respectively. On the one hand, based on the design of Sections 5.2 and 5.3, the experiments are conducted to determine the configuration schemes of activation function, loss function, and optimizer of the value function in the DQN algorithm, and verify the reasonableness and feasibility of the DQN algorithm in solving the task assignment problem of imaging satellites through the convergence, generalization, training efficiency, and testing performance of the algorithm; on the other hand, 6 learning-based bilevel task planning algorithms by integrating the deterministic algorithm and reinforcement learning are proposed through analyzing the performance of the algorithms in each integrating strategy, the effectiveness and superiority of the two algorithms, which are integrated HADRT with DQN, and integrated DPTS with DQN in solving the imaging satellite task planning problem are proved. Finally, the efficiency and solving effect of the algorithms in the training and testing processes are verified in the simulation scenarios. Then the application rules of the two integrated algorithms for different scenarios are reached.

The experiments were coded in Python. All experiments are implemented and compared on a laptop with Intel Core i7-8750H CPU @ 2.20 GHz, 16 GB RAM, and NVIDIA GeForce GTX 1060.

5.4.1 Tasking algorithm performance analysis

Based on the analysis in Section 3.5.2, it is not easy to find suitable indicators to evaluate the quality of results of task assignment directly. Relying on the task planning problem, this experiment analyzes the performance of DQN in solving the task assignment problem by controlling the algorithm and parameter configuration in the task scheduling phase.

(1) Experimental preparation

The preparation of experiments in this section mainly consists of two parts: the design of simulation scenarios and algorithm parameters.

1) *Experimental scenario design*

Table 5.1 shows the scenario-related parameters, satellite capability parameters, and main constraints of simulation scenarios.

Table 5.1: Experimental scenario design.

Classification	Detail design
Environment related	① Planning period is 24 hours; ② Time are discretized in units of 1 s; ③ Orbital elements of the satellite are same as Chapter 4.
Satellite capability	① The range of pitch angle and roll angle are $[-45°, 45°]$; ② The power and other energy resources of the imaging satellite at any time are sufficient, and energy-related constraints need not be considered; ③ Imaging satellites have limited memory storage capacity, which is capped at 750 TB; ④ Satellite maneuvering capabilities are limited, and it takes time to image one target and then turn to another.
Main constraints	① Each task can be executed at most once; ② Any task can only be performed within its visible time window; ③ The execution time window of any two tasks does not overlap; ④ Time interval between two consecutive tasks is not less than the minimum attitude transition time, which is determined by equation (4.8) and equation (4.9) together; ⑤ The sum of the storage consumed by all scheduled tasks is not greater than the storage capacity of the resource.

The pretreatment part of this experiment is the same as that in Chapter 4, which will not be repeated here. The objective of the task assignment problem considered in this section of the experiment is to maximize the total profit of the final task planning scheme. The specific objective function is as in equation (3.12), and the constraints are listed in Table 5.1.

Two types of distribution of tasks are designed in this experiment to verify the performance of algorithms: Chinese and Global.

Tasks in the "Chinese region" are randomly located in the regions 3°N to 53°N, 73°E to 133°E. Tasks in the "Global region" are randomly distributed in 65°S and 65°N, 180°W to 180°E in the region. The "Chinese region" scenarios test the efficiency of the algorithm when the tasks are distributed centrally, while the "Global region" scenarios test the performance of the algorithm when the tasks are distributed uniformly around the world. Experiments are conducted to test the scenarios of "Chinese region" with task sizes ranging from 100 to 400 and "Global region" with task sizes ranging from 100 to 600, and the proposed algorithms are compared with other algorithms.

In addition to the geographical location of each task, imaging duration, and profit are also necessary for describing a task. For most imaging satellites, the imaging duration is proportional to the area covered by the user's request. Apart from a few mapping imaging satellites take the imaging duration as the decision variable, imaging duration of each task on most imaging satellites cannot be changed. Assume that all users' requests are point targets, which means the imaging duration is a fixed number only related to the hardware design of the satellite. We set the imaging duration of each task as 5 s in the experiments.

The profit of each task is a number given by the user who submitted the corresponding request to the AEOS operation center. In order to standardize the description of the profit, it is usually discretized into 10 levels to describe the importance of a task, taking an integer ranging from 1 to 10. In the experiments, the value of the profit for each task is assigned randomly within this range.

2) *Design of parameters of the algorithm*

In the DQN algorithm, a neural network is used to characterize the value function in the MDP model. The scale of the parameters of the value function in this algorithm is related to the size of the task in the scenario; see Table 5.2.

Other necessary parameters in the algorithm are summarized as follows:
① Discount rate γ in training the Q-network: 0.9.
② Exploration rate ε: 0.2.
③ Topology of value network: fully connected.
④ Number of neurons in the input layer of the Q-network: $5n$.
⑤ Number of neurons in the hidden layer of the Q-network: 100.
⑥ Number of neurons in the output layer of the Q-network: $n + 1$.
⑦ Maximum number of experiences m in store: 10000.
⑧ Number of experiences b for training per batch: 100.

Table 5.2: Parameters in Q-network.

Scenarios	Input Layer	Hidden Layer	Output Layer	Sum
Chinese region_100	50100	10100	10201	70401
Chinese region_200	100100	10100	20301	130501
Chinese region_300	150100	10100	30401	190601
Chinese region_400	200100	10100	40501	250701
Global regions_100	50100	10100	10201	70401
Global region_200	100100	10100	20301	130501
Global region_300	150100	10100	30401	190601
Global region_400	200100	10100	40501	250701
Global region_500	250100	10100	50601	310801
Global region_600	300100	10100	60701	370901

⑨ Number of iterations e in each scenes: 20.
⑩ Number of training scenes c: 20.

(2) Configurations of value function

In order to discuss the influence of activation function, loss function, and optimizer on the performance of the algorithm, the experiment is carried out. Applying different configuration of functions to a simple scenario where the number of tasks is 20 and the total profit is 105. Trains 100 generations by DQN and record the total profit of the result. The detailed experimental results are shown in Figure 5.9 to Figure 5.11.

Figure 5.9, Figure 5.10, and Figure 5.11 contain three subfigures, respectively. Each subfigure represents the algorithm's convergence with different loss functions under the combination of an activation function and an optimizer. In general, no matter which optimizer or loss function is selected, Tanh activation function got the most unstable results, and no instance with Tanh can stably converge to the optimal solution (i. e., get the result of total profit 105 and 20 tasks); when Sigmoid activation function and SGD optimizer are selected, the results show two extremes: one can converge well to the optimal solution, and the other cannot converge within 100 generations. Under the ReLU activation function, the discrimination between different combinations of loss function and optimizer is relatively high, but has no significant regularity.

Table 5.3 shows the average profits obtained from training 100 generations under different activation, loss, and optimizer combinations. By analyzing the data in Table 5.3, the following conclusions can be drawn intuitively:

① Six of the 10 candidate loss functions show that ReLU outperformed other activation functions. This indicates that the DQN algorithm for solving the task assignment problem has better convergence when ReLU is chosen as the activation function in the value function.

② Among the above six loss functions, categorical cross-entropy (CC) performs better than other loss functions in 8 combinations of activation function and optimizer.

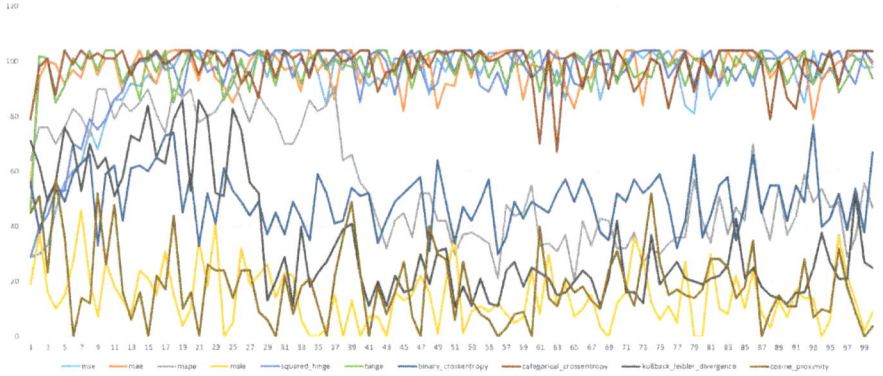

(a) SGD optimizer

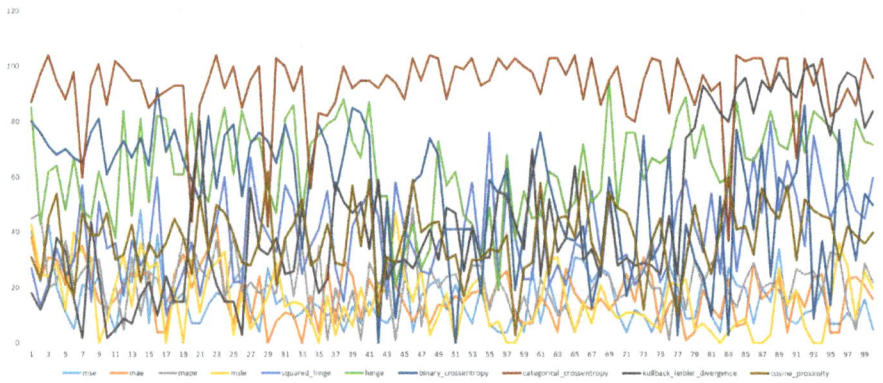

(b) RMSprop optimizer

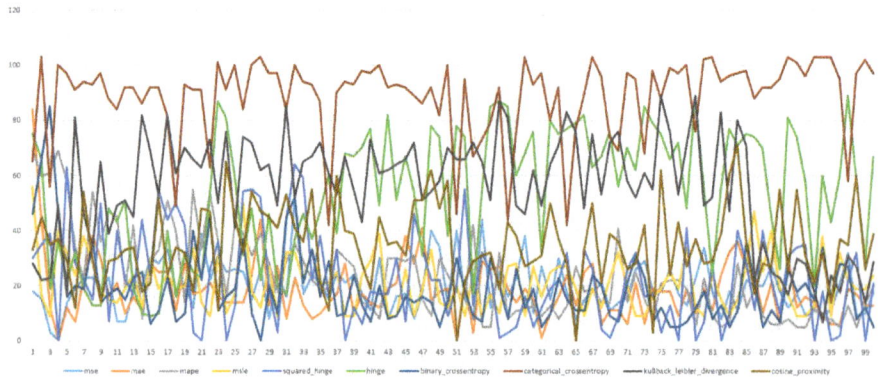

(c) Adam optimizer

Figure 5.9: Results of the algorithm with different loss functions and optimizers under the Sigmoid activation function.

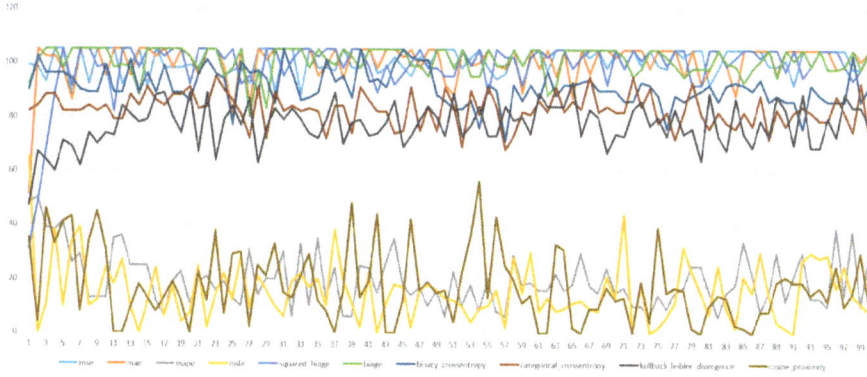

(a) SGD optimizer

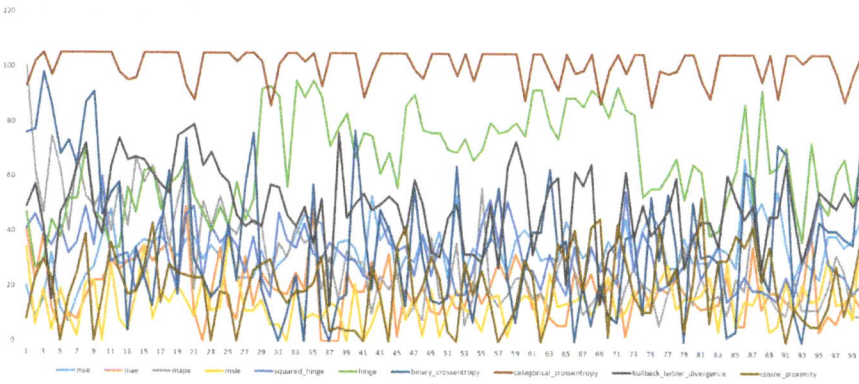

(b) RMSprop optimizer

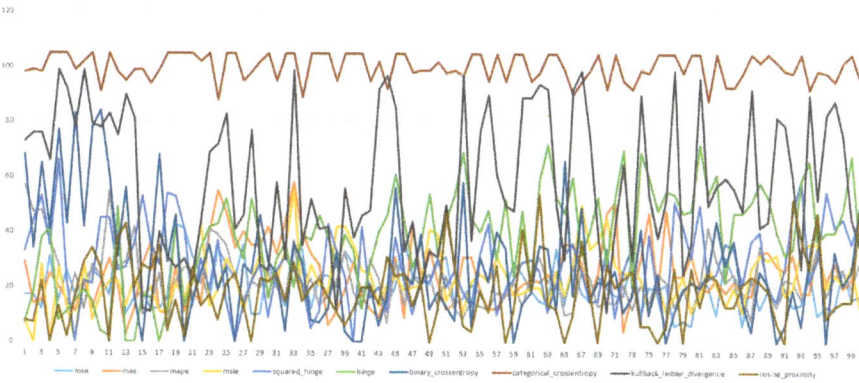

(c) Adam optimizer

Figure 5.10: Results of the algorithm with different loss functions and optimizers under the ReLU activation function.

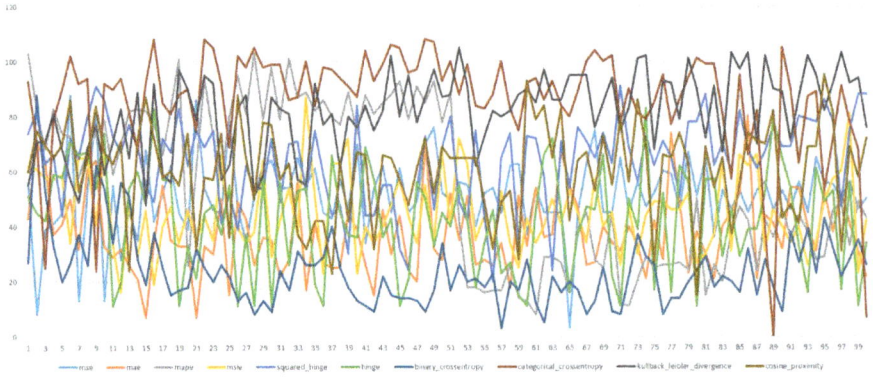

(a) SGD optimizer

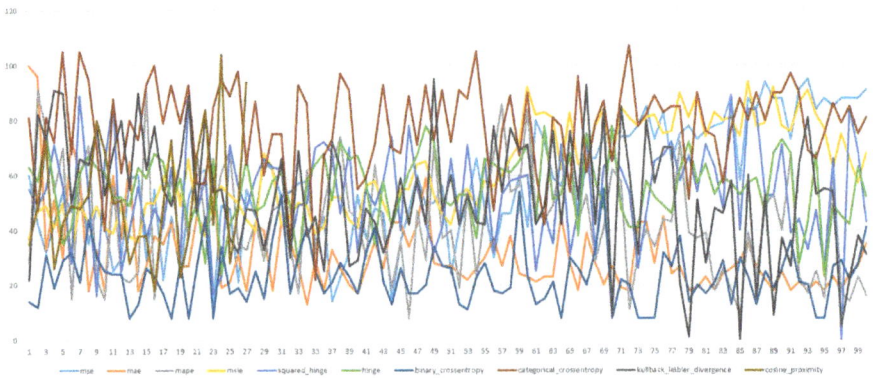

(b) RMSprop optimizer

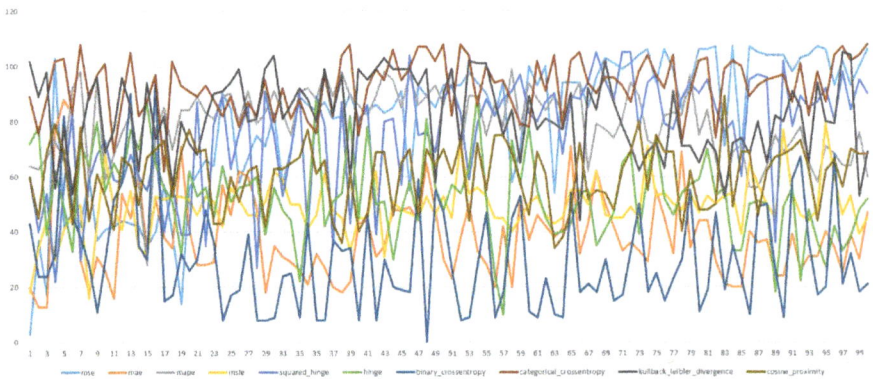

(c) Adam optimizer

Figure 5.11: Results of the algorithm with different loss functions and optimizers under the Tanh activation function.

Table 5.3: Average total profits of different value functions (100 iterations in single scenario).

Profit	Sigmoid			ReLU			Tanh		
	SGD	RMSprop	Adam	SGD	RMSprop	Adam	SGD	RMSprop	Adam
MSE	93.86	15.55	19.69	**100.7**	33.48	20.36	51.34	57.72	79.85
MAE	97.5	17.26	18.45	**100.65**	18.72	25.07	37.65	31.54	38.25
MAPE	56.61	21.52	23.73	19.7	30.42	22.54	55.6	42.74	**79.31**
MSLE	14.44	15.36	21.09	15.4	14.31	22.87	45.37	**61.74**	49.06
SH	94.49	39.4	22.77	**99.13**	31.18	28.3	64.15	53.59	75.49
HG	97.6	63.56	53.87	**100.91**	65.65	38.24	42.26	55.41	52.84
BC	49.76	53.77	18.91	**91.15**	35.75	27.26	23.81	23.12	29.45
CC	<u>98.63</u>	<u>92.71</u>	<u>88.81</u>	82.75	**101.17**	<u>100.21</u>	<u>86.33</u>	77.91	<u>93.04</u>
KLD	34.63	45.16	55.54	77.26	49.39	58.86	81.67	52.8	**82.11**
CP	18.51	37.71	35.19	17.4	20.35	18.67	**60.86**	51.74	60.4

Still, it performs poorly in the combination of SGD and ReLU, and is surpassed by the other five loss functions.

③ Overall, the combination of ReLU and CC performs the best among optimizers. Combined with the ReLU activation function, categorical cross-entropy (CC), and RMSprop optimizer, it can achieve the maximum average return among all 90 combinations.

Table 5.4 shows the standard deviation of the total benefits obtained from training 100 generations under different activation functions, loss functions, and optimizer configuration schemes. The standard deviation quantitatively describes the degree of dispersion of the sample data. The larger the standard deviation, the more divergent the data is considered. Therefore, in general problems, it is often hoped that the standard devi-

Table 5.4: Standard deviation of different value functions (100 iterations in single scenario).

Standard deviation	Sigmoid			ReLU			Tanh		
	SGD	RMSprop	Adam	SGD	RMSprop	Adam	SGD	RMSprop	Adam
MSE	15.62	9.32	<u>9.11</u>	**4.77**	8.17	9.58	13.95	20.77	24.45
MAE	**7.52**	<u>9.19</u>	11.27	7.53	11.56	10.32	14.38	15.20	15.20
MAPE	21.22	9.97	15.71	**9.89**	10.00	18.29	28.42	19.77	12.88
MSLE	10.60	11.21	10.11	11.03	10.11	**8.73**	14.12	16.44	<u>9.76</u>
sh	13.90	15.58	17.01	10.58	14.01	**10.20**	15.66	17.78	19.92
hg	7.41	16.56	23.00	**5.08**	17.79	17.50	19.10	12.20	15.56
BC	10.13	22.65	13.49	**6.42**	20.05	24.51	<u>12.96</u>	<u>10.37</u>	18.88
CC	<u>7.36</u>	12.66	14.41	5.97	**4.92**	<u>5.60</u>	19.37	16.02	10.14
kld	21.90	28.04	19.71	**7.93**	25.45	13.93	14.18	21.60	14.24
cp	13.45	37.71	14.37	13.29	18.67	12.89	16.67	51.74	**11.80**

ation is as tiny as possible. The following conclusions can be drawn from the data in Table 5.4:

① Seven of the 10 candidate loss functions show that ReLU can get the most stable results, which refers to the smallest standard deviation of the corresponding row.

② Categorical cross-entropy (CC) has better stability than other loss functions in most combinations. When choosing the loss function (CC), the minimum standard deviation of the corresponding column is obtained by three combinations with activation functions and optimizers: Sigmoid and SGD, ReLU and RMSprop, ReLU and Adam.

③ There are only two combinations with a standard deviation of less than five: "ReLU+SGD+MSE" and "ReLU+RMSprop+CC." The standard deviation of these two combinations is much smaller than that of others, but there is only a 0.15 difference. Therefore, it can be concluded that the stability of these two combinations is high.

Based on the above analysis, it can be concluded that selecting the activation function ReLU, loss function CC, and optimizer RMSprop can achieve the most substantial convergence and superior stability. Therefore, this combination will be adopted for all value functions in the subsequent experiments of this chapter.

(3) Results in training process
The performance of DQN in the training process is analyzed in this section, including convergence performance and training time of the algorithm.

1) *Convergence analysis*

In this section, the convergence of the proposed RL is tested. Figure 5.12 records the total profit of each episode in different scenarios, with 20 training scenarios and 20 iterations in each scenario. Each action will generate a data record about the action, state, and reward. These data are stored and recorded for training value function, i. e., Q-network. In each scenario, an action sequence will be formed to represent the task assignment scheme, and the total profit will be obtained based on the complete scheme. The total profit of each episode rises sharply at the beginning of training and then fluctuates around a certain value until the end of the training, even if the task set for training has changed. This shows that the trained Q-network is good at convergency and can be well applied to unknown scenarios.

The reason for the fluctuation in the total profit of Figure 5.12 is summarized as follows: The main reason is that the "Exploration" in the algorithm. The purpose of exploration is to enable the agent to find better solutions, but sometimes the quality of the solution may be worse than expected. When other conditions are unchanged, the larger the exploration rate, and the greater the fluctuation.

The total profit of all scenes converge in the long run, but the performances are different for every time at the beginning of training. Figure 5.13 shows the training data of the first 20 episodes in all scenarios. We saw a waving increment of the total profit in most of the scenarios, while the data in Global region_500, Global region_600, and

Figure 5.12: Line chart of total profit change over 400 iterations under different scenarios.

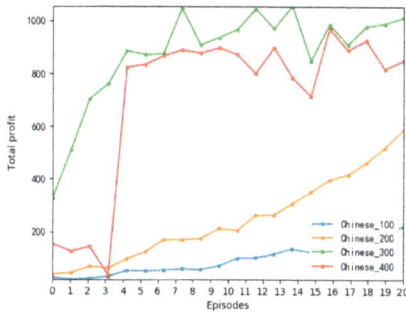

Figure 5.13: Trend of total profit in the first 20 generations of the algorithm in all scenarios.

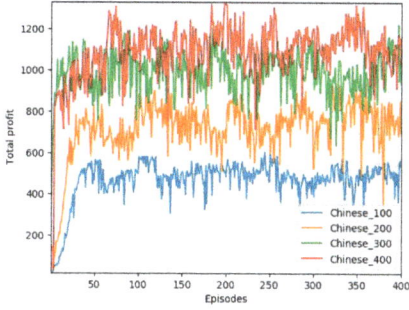

Chinese region_400 change dramatically at the first ten episodes. This is because when learning using a small number of samples, the algorithm may find a wrong value function, which leads to a wrong direction of the decision in the next one or several episodes. Nevertheless, the algorithm can always find the direction in which the Q-network converges from more experiences.

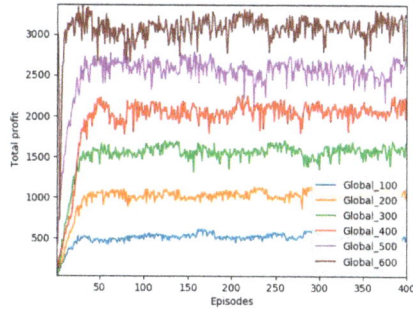

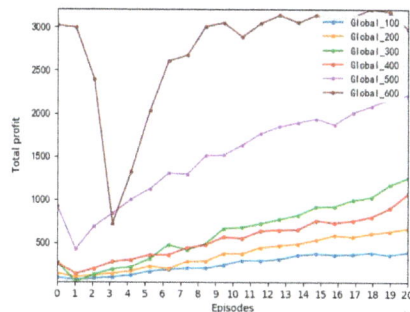

By synthesizing the performance of the algorithm in the training process in each scene, the convergence of the proposed DQN in the same scene and the generalization in different scenes can meet the application needs.

2) *Training time analysis*

The training time for each scenario is discussed in this section. According to the design in Figure 2.4, the simulation experiment starts from pretreatment, i. e., the algorithm reads the original information of the experiment, and prepares data by calling the pretreatment algorithm, so as to obtain the task information required for task planning. Due to the process of pretreatment involves some physical principles, such as orbital dynamics, the calculation process is complicated, and takes a long time. The experiment does not reform and design the internal of pretreatment, but only takes it as a part of the black box model in the whole process. This is to be close to the real-world engineering problems, simulate the complex and indefinable process of task planning in engineering, and finally discusses the algorithm. Therefore, the running time of the simulation includes the time required for preprocess, i. e., the time required to calculate the visible time window of each task and other necessary parameters for task planning by users' requirements, satellite capabilities, and orbital elements.

Table 5.5 shows the time spent on pretreatment and the learning process. From the results, the time spent on pretreatment is only related to the size of the task set. However, the time spent on learning is task-size-related and distribution-related. Both of them are positively correlated with task size, i. e., the larger the task size, the longer the time required for training and pretreatment. The longest time spent on training in these scenarios we design appears in Chinese region_400, which is about 3 h and 45 min. It is fully acceptable to take several hours to train a value function for an imaging satellite task assignment model.

Table 5.5: Time consumed by training process.

Scenarios	Pretreatment time	Learning time	Time spent on each data set (per 20 episodes)
Chinese region_100	21 m 52 s	5 m 11 s	15 s
Chinese region_200	44 m 28 s	27 m 43 s	1 m 23 s
Chinese region_300	1 h 6 m 54 s	1 h 8 m 4 s	3 m 24 s
Chinese region_400	1 h 30 m 14 s	2 h 15 m 6 s	6 m 45 s
Global region_100	22 m 38 s	4 m 0 s	12 s
Global region_200	45 m 10 s	10 m 12 s	30 s
Global region_300	1 h 7 m 52 s	17 m 23 s	52 s
Global region_400	1 h 30 m 32 s	28 m 6 s	1 m 24 s
Global region_500	1 h 51 m 25 s	38 m 44 s	2 m 2 s
Global region_600	2 h 12 m 49 s	50 m 27 s	2 m 31 s

(4) Results in testing process

In order to test the application effect of the value function trained by the DQN in unknown scenes, several advanced imaging satellite task planning algorithms are selected for experimental comparison: ALNS and HADRT. The branch-and-bound algorithm is also considered [133], but it cannot solve the problem in 3600 s, which is intolerable for decision-makers in most real-world applications. We compare and discuss the results of ALNS, HADRT, and the RL approach proposed in this work. It is worth mentioning that HADRT is a constructive heuristic algorithm proposed in Chapter 4. It can be nested in the environment of reinforcement learning algorithm as a calculation method for task scheduling process, or applied directly for solving imaging satellite task planning problems. Therefore, comparing the task planning algorithm integrated with HADRT and

DQN algorithm with simple HADRT, the contribution rate of the DQN to improve the quality of the solution can be dug out. A same experiment data set is selected to test the performance of different algorithms in the test scenario.

1) *Comparison of the rate of total profit*

The rate of total profit is one of the most important indexes for evaluating imaging satellite planning algorithms, which indicates the quality of a scheme. The objective of the problem is to maximize the total profit of the scheme. The rate of total profit is proportional to the total profit in the same task set. It can be calculated by equation (4.37).

ALNS, HADRT, and RL have been tested in ten different scenarios, and we have tested ten times under different task sets on each scenario to reduce the influence of contingency. The profit rate of these three algorithms in different task sets are shown in Figure 5.14.

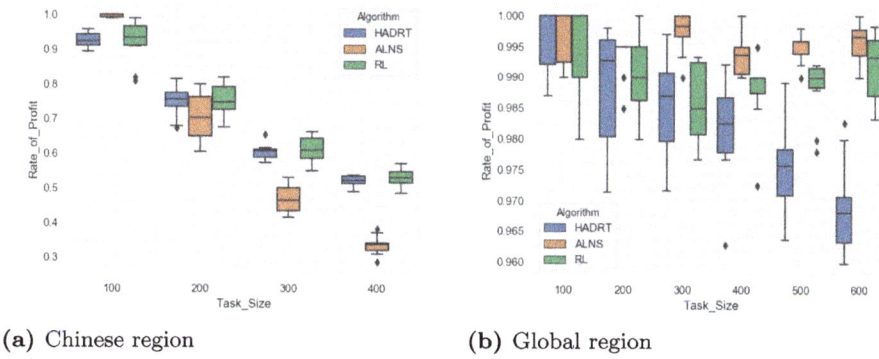

(a) Chinese region   (b) Global region

Figure 5.14: Profit rate of three methods in different test sets.

Figure 5.14a and Figure 5.14b are box diagrams to illustrate the difference in profit rate in various cases. We called scenarios in which most tasks can be scheduled (more than 90 %) as undersubscribed cases; otherwise, the scenario is oversubscribed. It is easy to learn from Figure 5.14 that most cases in Chinese region_100 and all scenarios in global distribution are under-subscribed, while all cases in Chinese region_200, Chinese region_300, and Chinese region_400 are oversubscribed. Most researches on imaging satellite task planning only discuss the performance of the algorithm in the case of oversubscribed, but in practical applications the performance of the algorithm in the case of undersubscribed also has significant practical meaning.

The average profit rate of HADRT is superior to that of ALNS in cases of oversubscribed, while the conclusion is reversed in cases of undersubscribed. The gap between results of the two algorithms are becoming larger with the task size growing. The results of different data sets are opposite, so neither ALNS nor HADRT can guarantee the stable high performance in most of scenarios. It is worth noting that:

① The average profit rate of DQN exceeds that of HADRT in all test scenarios, which indicates that the quality of the solution can be improved by DQN.
② DQN gets the highest average profit rate in all of the oversubscribed scenarios and scenario Global region_400. In other scenarios, although the average profit rate of DQN does not exceed ALNS, there is only a marginal difference between the two values: The relative error between the average profit rate of DQN, ALNS, and HADRT in the same scenario is less than 0.5 %.

To sum up, the following conclusions can be drawn:
① ALNS performs the worst in oversubscribed scenarios, DQN performs the best in oversubscribed scenarios.
② In scenarios of undersubscribed, HADRT performed the worst; while, DQN and ALNS have no absolute advantage in solving accuracy in this type of scenarios.

2) *Comparison on CPU time*

In order to discuss the performance of DQN comprehensively, the experiment compares the CPU time of algorithms, as shown in Figure 5.15.

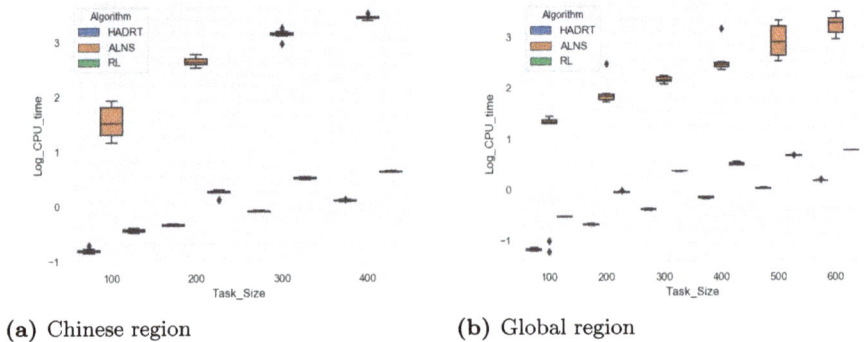

(a) Chinese region   (b) Global region

Figure 5.15: CPU time of three methods in different test sets (logarithmic).

CPU time is another crucial index to evaluate an algorithm. CPU time of DQN refers to the time consumed used in the testing process. The following conclusions can be drawn from Figure 5.15:
① The CPU time of HADRT and RL are both at a low level and stable in all scenarios. In the same scenario, DQN takes slightly more time than HADRT, which is the time required for task assignment. Therefore, the task assignment process based on DQN is completely acceptable in terms of time efficiency.
② ALNS is always the most time consuming algorithm, specifically, its time consumption is generally 2 to 3 orders of magnitude higher than the other two algorithms. Its average CPU time for more than 40 s in Chinese region_100. In Chinese region_200, the value increases ten times of that in the condition of Chinese region_100, while

it becomes more than 2700 s in the scenario of Chinese region_400. Meanwhile, a similar trend is spotted in global scenarios: CPU time of ALNS grows significantly with the size of task sets.

According to Table 5.6, the rule of CPU time of each algorithm can be quantitatively analyzed. By comparing the Chinese and global regional scenarios under the same task number, it is found that the distribution of tasks will also lead to the difference in CPU time of the algorithm. It is also noticeable that the standard deviation of ALNS is much larger than that of RL and HADRT, for ALNS is a type of stochastic algorithm, and the algorithm will stop iteration after reaching the convergence condition or the maximum number of iterations, so it is hard to estimate the calculation time. In a word, compared with RL and HADRT, ALNS performs the worst in time efficiency and stability.

Table 5.6: Statistics of the average and standard deviation of CPU time.

Scene	Average CPU time (s)			Standard deviation		
	ALNS	HADRT	RL	ALNS	HADRT	RL
Chinese region_100	4.37E+01	1.57E−01	3.61E−01	2.55E+01	1.61E−02	2.43E−02
Chinese region_200	4.46E+02	4.56E−01	1.77E+00	7.15E+01	1.70E−02	2.22E−01
Chinese region_300	1.39E+03	8.18E−01	3.25E+00	2.11E+02	1.63E−02	1.14E−01
Chinese region_400	2.75E+03	1.27E+00	4.21E+00	2.11E+02	4.01E−02	1.02E−01
Global region_100	1.80E+01	6.68E−02	2.95E−01	9.27E+00	2.57E−03	3.96E−03
Global region_200	8.74E+01	2.07E−01	8.97E−01	6.95E+01	7.71E−03	2.29E−02
Global region_300	1.45E+02	4.12E−01	2.28E+00	1.79E+01	1.25E−02	2.59E−02
Global region_400	3.87E+02	6.94E−01	3.13E+00	3.43E+02	2.22E−02	2.21E−01
Global region_500	9.98E+02	1.04E+00	4.53E+00	6.30E+02	2.61E−02	7.29E−02
Global region_600	1.81E+03	1.47E+00	5.87E+00	6.95E+02	2.71E−02	2.66E−02

The above analyses of the profit rate and the CPU time for different algorithms, the conclusions can be summarized as follow: HADRT performs best in terms of CPU time, but the total profit in most scenarios is less than the other two algorithms; ALNS achieves high total profit in some scenarios, but CPU time is unacceptable. Considering the total profit and CPU time comprehensively, DQN can get a satisfactory solution in an acceptable time.

5.4.2 Performance of integration algorithms

(1) Experiment preparation

Resource settings for the experiment are the same as those in Section 5.4.1. See Table 5.1 for parameters. These parameters are designed according to the real data provided by the satellite industry department and meet the standards of the satellite industry.

Due to the complexity of the real-world problems and the diversity of inputs, it is difficult to find a universally recognized and practical benchmark to study the imaging satellite task planning problem. In order to facilitate comparison and analysis, [14, 73, 100] conduct experiments by simulation. The test data set used in our experiment is designed referring to literatures. Table 5.7 lists the rules for generating tasks in simulation scenarios.

Table 5.7: Task generation rules in simulation scenarios.

Parameters	Distribution	Type	Lower bound	Upper bound
lat_i in large-area scenes	Evenly distributed	Float	3	53
lon_i in large-area scenes	Evenly distributed	Float	73	133
lat_i in small-area scenes	Evenly distributed	Float	20	30
lon_i in small-area scenes	Evenly distributed	Float	108	114
d_{i}^{j}	Constant	Int	5	5
p_{i}^{j}	Evenly	Int	1	10

Two types of scenarios were designed to test the effectiveness of the proposed method in different geographical distributions of tasks. The ability to handle routine tasks could be tested in large-area scenes, which covers the whole are of China; the responsiveness of AEOS in emergencies like floods and earthquakes could be tested in small area scenes, which covers the Hunan province in China. Accurately, two small area scenes with task sizes of 20 and 50 as well as three large area scenes with task sizes of 100, 200, and 400 were designated, which are marked as H_20, H_50, C_100, C_200, and C_400, respectively.

(2) Integrated algorithms

In the MDP model established in Section 5.2, the calculation of the environment is mainly realized by task scheduling algorithms. Before analyzing the performance of the integrated algorithm, the experiment discusses the convergence of the integrated algorithm formed by scheduling algorithms and assignment algorithms in imaging satellite task planning problems.

Asynchronous advantage actor critic [179] (A3C) and actor critic algorithm with pointer-networks [102] (PtrN) are selected as comparison algorithms of DQN. These two algorithms have shown good performance in planning and scheduling problems in literatures recently. For controlling the variables, the same hyperparameters are applied to these three algorithms, including the learning rate, exploration rate, and the structure and the hyperparameters of the value network. Combining the two deterministic algorithms designed in Chapter 4, six algorithm integration schemes are constructed, as shown in Table 5.8.

Table 5.8: Six integrated algorithms.

No.	Task assignment algorithms	Task scheduling algorithms	Bilevel algorithms
1	DQN	DPTS	DQN_DP
2	DQN	HADRT	DQN_CH
3	A3C	DPTS	A3C_DP
4	A3C	HADRT	A3C_CH
5	PtrN	DPTS	PtrN_DP
6	PtrN	HADRT	PtrN_CH

The scenario C_400 is chosen for discussing the convergence of algorithms, so as to increase the discrimination between the results of different methods, making it easier to analyze the results. We run these six algorithms 1000 times in the same scenario, and the total profit recorded in each iteration is displayed in Figure 5.16.

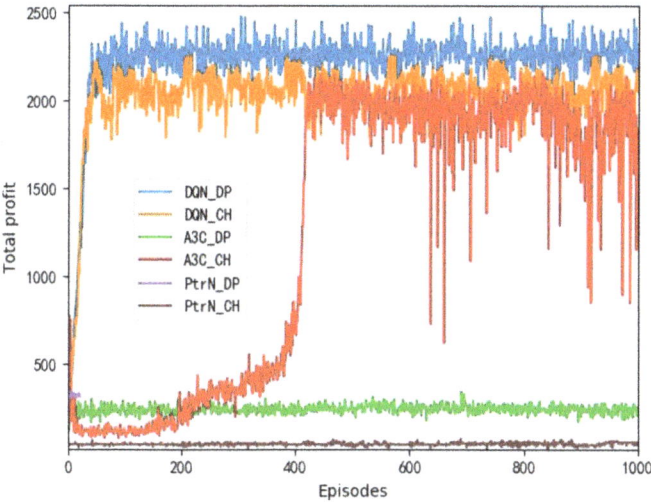

Figure 5.16: Convergence of different integrated algorithms.

It can be found from Figure 5.16.

① DQN_DP, DQN_CH, and A3C_CH can get high total profits before 1000 episodes, Others maintains a low level of the total profit during the first 1000 episodes.

② DQN_DP and DQN_CH are algorithms with high convergence, and they have converged in the first 50 episodes. However, A3C_CH performs poor in convergence, but it increases sharply in total profit around 400 generations. In a word, it is not as stable as DQN_DP and DQN_CH.

③ Under limited computing resources (i7-8750H CPU, 16.0 GB RAM, GTX 1060), PtrN_DP can only complete 22 episodes of training due to memory overflow. This indicates that among all six integrated algorithms, the PtrN_DP algorithm requires the highest computational resources. The training and testing processes of other algorithms can be completed on a personal laptop.

④ The convergence accuracy of the six algorithms is ranked from high to low as DQN_DP, DQN_CH, A3C_CH, A3C_DP, and PtrN_CH. It is worth mentioning that based on the data from the previous 20 generations, the accuracy of PtrN_DP is higher than that of A3C_DP and PtrN_CH. However, as the number of episode increases, it cannot be determined based on this experiment how the results of PtrN_DP change.

Hence, among these six integrated algorithms mentioned above, DQN_DP and DQN_CH are algorithms with advantages in fast convergence speed, high solving accuracy, low requirements for computing resources, and better convergence level than the other four integrated algorithms. In the following experiments, we will discuss in detail the various performances of DQN_DP and DQN_CH during training and application.

(3) Results in training process

1) *Convergence analysis*

1000 iterations on the same instance are used to verify the convergence of the proposed DQN_DP and DQN_CH, as shown in Figure 5.17. Through dozens of iterations, the total profit of all algorithms is stable overall, with acceptable fluctuations. The fluctuation is mainly caused by the existence of "Exploration" in the training process of DQN.

There was no significant difference between the two algorithms on the training sets H_20 and H_50; see Figure 5.17a and Figure 5.17b, respectively. This indicates that the algorithm can achieve the same level of training accuracy in small-scale scenarios, and the computational resources consumed during the training process are relatively small. From Figure 5.17c, Figure 5.17d, and Figure 5.17e, it can be observed that as the number of tasks increases, although both algorithms can effectively converge, the differences between the two algorithms gradually become apparent: The total profit of DQN_DP averages around 10 % higher than that of DQN_CH, but DQN_DP requires more computing resources and time during the training process: it can only complete 949 iterations in C_400 because of memory overflow. It reveals that:

① Both DQN_DP and DQN_CH can train empirical formulas for solving task assignment problems with small-scale samples under limited computing resources.

② DQN_DP has higher training accuracy, but it requires more computing resources; The training accuracy of DQN_CH is slightly poor, but it requires low computational resources and can obtain satisfactory solutions in larger and more complex scenarios.

2) *Generalization analysis*

Twenty random instances in 20 random scenarios are used to test the generalization of the proposed methods. See the results in Figure 5.18. Depending on the different scales,

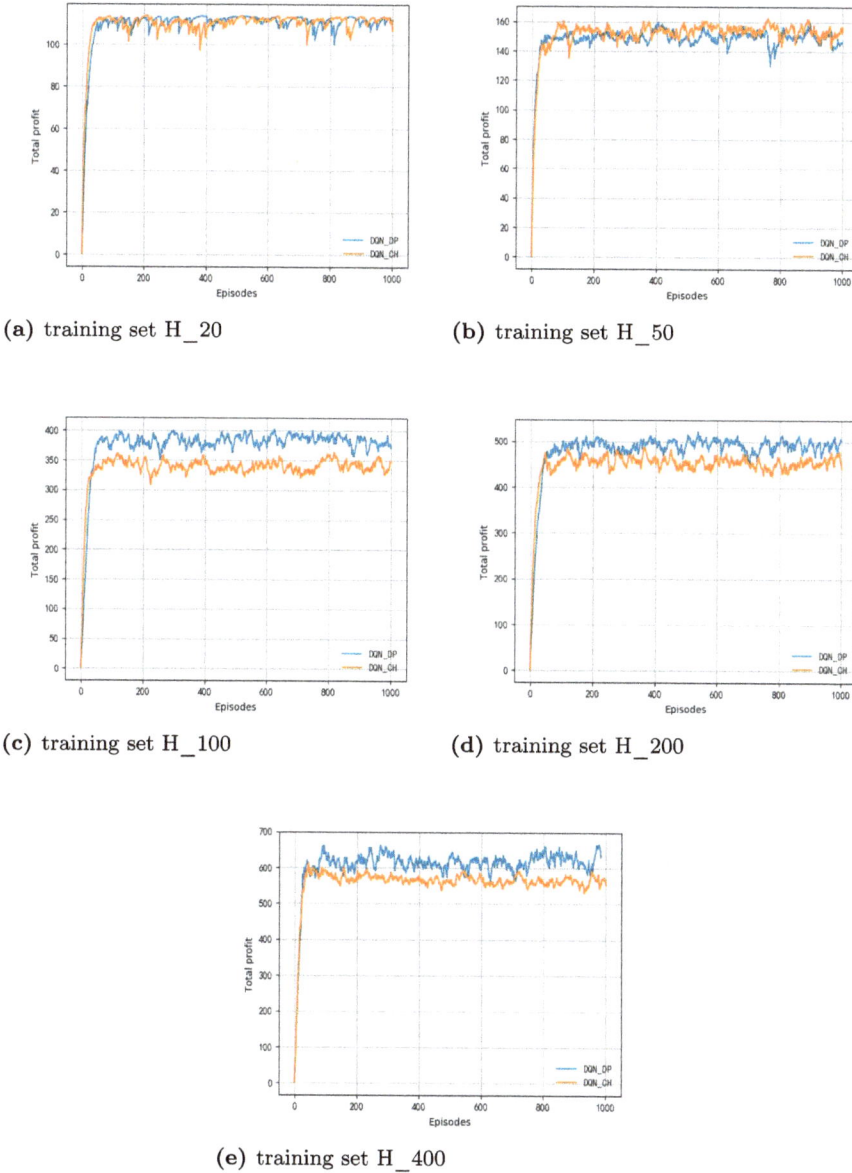

(a) training set H_20

(b) training set H_50

(c) training set H_100

(d) training set H_200

(e) training set H_400

Figure 5.17: Convergence analysis of DQN_DP and DQN_CH.

training processes of different instances take tens of minutes to several hours, including the time for pretreatment and status updates. Since the profit of each task in a task set is random, analyzing the trend of profit margin is more reasonable. The profit margin is the ratio of the total profit of scheduled tasks to that of all candidate tasks; see equation (4.37). As can be seen in Figure 5.18, although only 20 iterations are performed on each

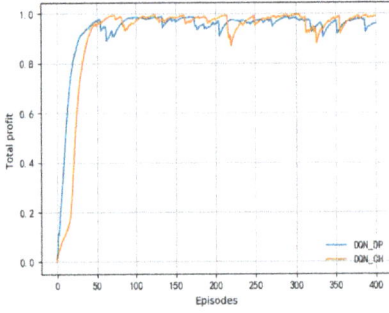

(a) training set H_20

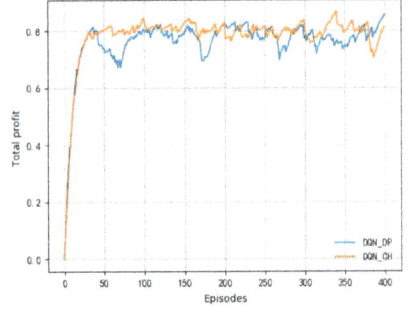

(b) training set H_50

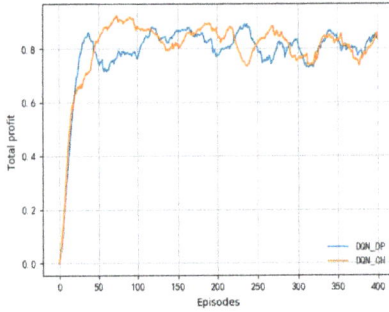

(c) training set H_100

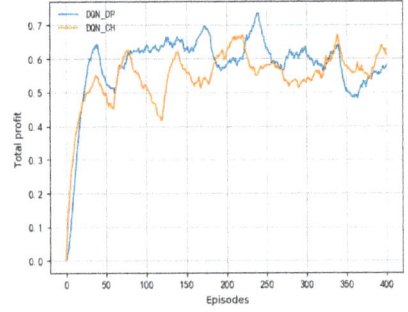

(d) training set H_200

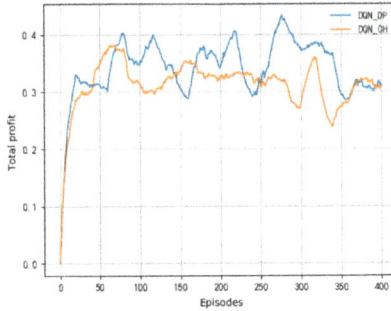

(e) training set H_400

Figure 5.18: Generalization analysis of DQN_DP and DQN_CH.

data set, the polyline of each algorithm fluctuates across a certain level, indicating the value functions trained by these two algorithms are applicable in different scenarios, i. e., DQN_DP and DQN_CH are robust in generalization. However, the fluctuations in Figure 5.18 are larger than the fluctuation range in Figure 5.17. This is because:

① The optimal solutions vary in each instance.

② The value function need to be continuously modified in different instances.

The value function trained in 20 training scenarios will be applied to practical application scenarios to further test the generalization ability of DQN_DP and DQN_CH. The results are shown in Figure 5.19. Among the 50 instances with different features in the figure, the test results of the two algorithms are basically consistent with the training performance: except for the instance in H_20, the results of DQN_CH and DQN_DP are only equal on one instance; among the 50 test samples, the DQN_CH algorithm performed better than the DQN_DP algorithm only in two samples.

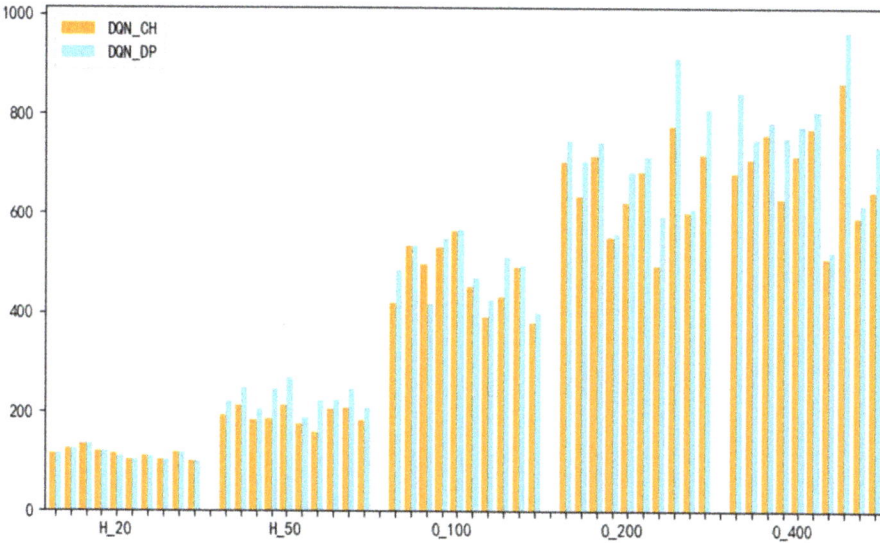

Figure 5.19: Experimental results of generalization in test sets.

(4) Testing results

Compare DQN_DP and DQN_CH with several advanced algorithms in published papers to obtain relevant conclusions on the application process of the proposed algorithms. These comparison algorithms include: exact algorithms represented by branch and bound algorithms [133], constructive heuristic algorithms represented by heuristic algorithm based on the density of residual tasks designed in Chapter 4, as well as metaheuristic algorithms represented by the adaptive large neighborhood search algorithm [14] and bidirectional dynamic programming based iterated local search [78]. These algorithms are represented by B&B, HADRT, ALNS, and BDP-ILS, respectively.

We ran concerned algorithms in 50 instances and recorded results whose running time less than 3600 seconds. All results are summarized in Figure 5.20. The bar chart shows:

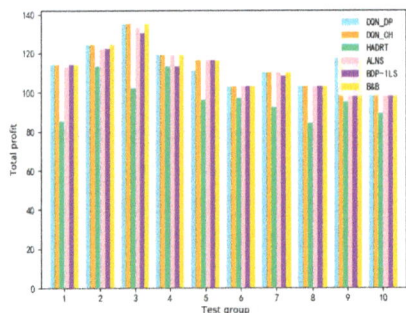

(a) H_20

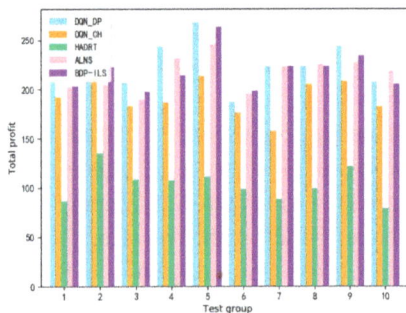

(b) H_50

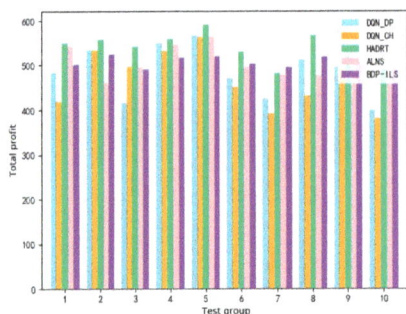

(c) H_100

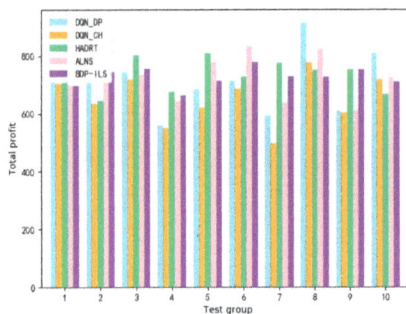

(d) H_200

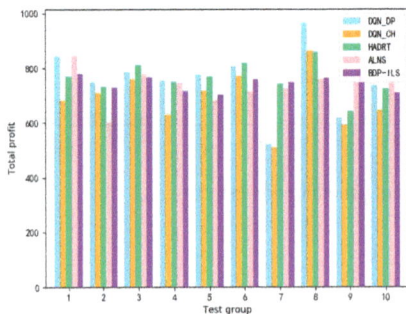

(e) H_400

Figure 5.20: Comparison on test results.

① B&B could obtain the optimal solution in small scale scenes (H_20), but cannot obtain the result within 3600 seconds in other test sets.
② HADRT reaches acceptable results in some instances of test set C_100, C_200, and C_400, but in others (especially H_20 and H_50), the results are not satisfactory.
③ BDP-ILS, DQN_CH, and DQN_DP reliably achieve high-quality results, and DQN_DP is superior to DQN_CH in most instances.

Running time is another important metric to evaluate the efficiency of algorithms. This is also significant index for the actual application efficiency and user experience. Table 5.9 records the average operation time of each algorithm on different data sets. Based on this table, the following conclusions can be drawn:

① HADRT, DQN_CH, and DQN_DP could get results in a short time, while the running time of ALNS, BDP-ILS, and B&B grow significantly with the increase of the problem size.

② The computation time of DQN_CH is approximately twice that of HADRT, while the running time of these two algorithms is several orders of magnitude smaller than that of other algorithms.

③ Although the running time of DQN_DP is much larger than that of DQN_CH and HADRT, the increase rate with the scale of the problem is acceptable.

Table 5.9: The average running time on different data set.

Scenarios	DQN_CH	DQN_DP	HADRT	ALNS	BDP-ILS	B&B
H_20	0.079	1.824	0.038	7.777	1.094	1.553
H_50	0.153	5.539	0.126	20.287	2.085	>3600
C_100	0.399	9.367	0.152	38.691	10.558	>3600
C_200	1.003	24.655	0.442	316.83	43.492	>3600
C_400	1.809	24.426	1.246	2057.9	180.29	>3600

By analyzing the standard deviation of running time in Table 5.10, it can be found that the standard deviation of ALNS and BDP-ILS is large, especially in scenarios C_200 and C_400. Under the same test group, the computation time of ALNS, BDP-ILS, and B&B is greatly affected by the characteristics of inputs. By contrast, HADRT, DQN_CH, and DQN_DP run with a stable time.

Table 5.10: The standard deviation of running time on different data set.

Scenarios	DQN_CH	DQN_DP	HADRT	ALNS	BDP-ILS	B&B
H_20	0.004	0.038	0.013	1.146	0.429	1.234
H_50	0.028	0.38	0.011	11.653	0.474	–
C_100	0.024	0.752	0.017	26.916	3.923	–
C_200	0.12	1.44	0.034	75.318	23.131	–
C_400	0.234	2.031	0.098	222.41	34.935	–

Overall, B&B only work well on limited instances. HADRT could find a feasible solution in a short time, but the quality of the solution could not be guaranteed. For ALNS and BDP-ILS, the risk of falling into a local optimum can be reduced through the random search mechanism, but it usually takes too long to search for a promising solution, which may be unacceptable in many practical problems. DQN_CH and DQN_DP obtain satisfactory solutions with acceptable time.

Among them, DQN_DP is superior in solving accuracy in most scenarios, while DQN_CH has more advantages in computing resource consumption. Therefore, when solving large-scale problems or computing resources are limited, selecting DQN_CH can get a satisfactory task planning scheme in a short time; when solving small-scale problems, selecting DQN_DP can further improve the quality of solution.

5.5 Summary

In this chapter, the imaging satellites task assignment process is modeled as MDP. In this model, the process of task assignment and task scheduling interact continuously to train the value function applied for assigning tasks. Assigning tasks is based on the value function learned by MDP while scheduling tasks is based on the deterministic algorithm designed in Chapter 4. Aiming at the problem of imaging satellite task assignment, the state, action, reward, and value function of the MDP model are designed to build an imaging satellite task assignment model that is as realistic as possible. The MDP model not only reflects the essential characteristics of imaging satellite task planning problems but also eliminates situations that are inconsistent with the real problem. Based on the understanding of the problem, an improved deep Q-learning (DQN) algorithm is proposed to solve the above MDP model. Combined with domain knowledge, it designs the solution framework of the algorithm and the action-pruning strategy at each decision-making step, which reduces the solution space and improves the convergence efficiency of the value function, thus improving the efficiency and accuracy of the final solution.

Experiments verify the performance of the improved DQN in the task assignment process and the performance of the learning bilevel task planning algorithm in the imaging satellite task planning problem. Through analysis of the value function in the DQN algorithm, the influence of loss functions, activation functions, and optimizers in the value function on the algorithm efficiency is studied. Based on this, it is proved the DQN can transfer the complex calculation process offline and achieve "fast and well" when assigning tasks. Furthermore, by combining the three advanced reinforcement learning algorithms with the two deterministic algorithms proposed in Chapter 4 of this book, six integrated algorithms can be constructed, and the impact of integrations on algorithm performance is studied. Results show that the integrated algorithms DQN_DP and DQN_CH are superior to the other four kinds of integrated algorithms in terms of convergence speed, accuracy, and stability. Among them, DQN_CH can obtain satisfactory convergence accuracy with minimal time cost and computational resources; DQN_DP can obtain higher convergence accuracy, but it needs more computing resources and time.

Furthermore, the integrated algorithms DQN_DP and DQN_CH will be applied to more complex engineering problems to verify the value of integrating deterministic algorithms and reinforcement learning to solve imaging satellite task planning problems in practice.

6 Application study on "SuperView-1" imaging satellites task planning

The technical characteristics of "SuperView-1" commercial remote sensing satellite constellation bring new challenges to the task planning process, such as the difficulties in unified modeling, constraints processing, and fast solving. Based on the background of the project, this chapter follows the framework and specifications of the current satellite ground operation and control system, designs the internal and external interfaces and data structure of the "SuperView-1" task planning system, establishes a bilevel optimization model for the actual problem, and realizes "SuperView-1" constellation learning-based bilevel task planning algorithms. Through simulation experiments of 14 daily task planning scenarios, the algorithm's solution accuracy, capacity boundary, satellite load balance, and other indicators are verified, and the effectiveness of the bilevel optimization model and integrated algorithms proposed in this book is proved in practical engineering problems.

6.1 Background for "SuperView-1" task planning

6.1.1 Basic information of "SuperView-1"

The successful launch of the "SuperView-1" commercial remote sensing satellite constellation marks the arrival of the era of independently developed commercial remote sensing in China [180]. The completion of this system can customize high-resolution remote sensing data according to user needs, providing robust information support in specific social activities such as land resources census, geographic surveying and mapping, environment monitoring, transportation, disaster prevention and mitigation, etc. [3]. According to Chinese long-term planning for the development of commercial remote sensing satellites, in the future, China will form a high-resolution commercial remote sensing constellation consisting of 16 high-resolution optical imaging satellites, four optical imaging satellites, four microwave imaging satellites, and several micro-satellites consisting of video imaging satellites and hyperspectral imaging satellites, i. e., the "16+4+4+X" system [181]. "SuperView-1 01" to "SuperView-1 04" are the first components of the "16+4+4+X" system, with four 0.5-meter resolution optical imaging satellites divided into two groups, which were successfully launched in two batches on December 28, 2016, and January 9, 2018, respectively [182]. Unless otherwise noted, the terms "'SuperView-1' commercial remote sensing satellite constellation" are abbreviated to "'SuperView-1' constellation" in the subsequent contents of this chapter, "'SuperView-1' commercial remote sensing satellite and constellation task planning process/project/model/problem/algorithm" are shortened to "'SuperView-1' task planning process/project/model/problem/Problem/Algorithm."

https://doi.org/10.1515/9783111585109-006

From the perspective of satellite platform and payload design, "SuperView-1" constellation has four major technical characteristics:

(1) High imaging resolution

The image resolution of the four satellites in the "SuperView-1" constellation is better than 0.5 meters [6], a level achieved only by the United States, South Korea, and a few other countries. Enhancing the image resolution of imaging satellites is one of the important indicators for improving the level of satellite services. Satellite industry departments around the world will continue to pursue high image resolution of imaging satellites because the breakthrough in resolution of imaging satellites will give rise to more and more remote sensing application modes: through the 0.5-meter panchromatic resolution image, the number of sewer entrances and exits within a city street can be counted, the contours of a car can be identified, and the more subtle geologic changes can be distinguished, and so on. Compared with low-resolution imaging satellites and constellations, the "SuperView-1" constellation dramatically improves the application scope of satellites, and it can extract a large amount of information that is difficult to mine from low-resolution satellite images according to specific needs. As a result, the imaging requirements of the "SuperView-1" constellation are increasing daily, which puts higher requirements on the computational efficiency of the satellite task planning method.

(2) Advanced imaging mode

Each of the "SuperView-1 01" to "SuperView-1 04" carries a panchromatic, multispectral imaging payload with a maximum resolution of 0.5 m. When the payload is in operation, the imaging modes of the payload can be adjusted according to different needs to meet the user's product requirements on the premise of saving as much power and storage space as possible when imaging satellites. The imaging payload carried by "SuperView-1" satellites embedded four imaging modes: multispectral nondestructive imaging, multispectral 2:1 compressed imaging, panchromatic 2:1 compressed imaging, and panchromatic 4:1 compressed imaging. The satellite power and storage consumption varies when working on different imaging modes. It leads to diverse task requirements, increasing the difficulty of modeling the task planning problem in a unified manner.

(3) Composite work mode

According to the user's application requirements for satellite images, "SuperView-1" constellation can realize a variety of task modes: single maneuvering imaging, continuous maneuvering imaging, real-time transmission, single-antenna data transmission, dual-antenna data transmission, and calibration. Through the flexible application of these task modes, a variety of real-world needs can be met: synthesizing three-dimensional images by shooting the same target from different angles, synthesizing regional images by dividing an imaging region, and forming multiple imaging tasks to be executed continuously in a short period of time. Different task modes impact resource usage differently in the model, corresponding to different constraints. It leads to a diversity of constraints in the model, and an efficient and generalized way of constraint handling needs to be found.

(4) Agility maneuverability

The strong attitude maneuvering capability of the "SuperView-1" constellation enables it to achieve faster response and broader coverage of imaging targets. Enhanced attitude maneuvering capability can alleviate the limitations of the satellite flight orbit on target visibility, allowing more imaging opportunities for imaging targets. This feature brings more freedom to the task planning process, and the algorithm can adjust the task planning scheme more flexibly to meet more users' imaging needs. On the other hand, this change leads to a steep increase in the solution space of the task planning model, and traversing all the nodes in the solution space becomes increasingly unacceptable in terms of time. It requires the algorithm to design a more scientific search strategy to improve the algorithm's solution quality in complex real-world scenarios.

The platform and payload design of the "SuperView-1" constellation aims at a high level of commercial imaging satellites, featuring high imaging precision, high agility, multimode, and other advancements. These features make "SuperView-1" satellites capable of carrying out complex tasks independently. However, due to the orbital and payload limitations, there is an apparent ceiling in the number of tasks and time efficiency of a single satellite. Collaborative task planning with multiple satellites can effectively broaden the boundaries of a single satellite's capabilities, generate new application modes, and significantly increase the overall application efficiency of the system.

Four satellites in the "SuperView-1" constellation are deployed on the same orbital plane, with a phase difference of 90 degrees between two neighboring satellites, for realizing rapid and accurate imaging for targets within a short period of time while taking into account the breadth of the coverage area. Through the coordinated planning of the constellation, it is possible to revisit any global target within one day. Also, it has the ability to collect over 3 million square kilometers of image data in one day [181]. Table 6.1 shows two line elements for the "SuperView-1" constellation (data retrieved on October 8, 2021, at 17:00 at https://celestrak.com/satcat/).

The powerful hardware capability has made satellite users and managers look forward to the practical application efficiency of the "SuperView-1" commercial remote sensing satellites. However, no matter how high the hardware level of imaging satellites develops, their ability to acquire images is always limited. When the scale of user demand exceeds the capacity boundary of the satellite system, the coordination of satellite resources through the design of reasonable task planning methods can effectively alleviate the contradiction between limited resources and growing user demand.

Table 6.1: TLE of "SuperView-1" constellation.

Label	Meaning	Satellite 01	Satellite 02	Satellite 03	Satellite 04
01	Line Number of Element Data	1	1	1	1
03-07	Satellite Number	41907	41908	43099	43100
08	Classification	U	U	U	U
10-11	International Designator (Last two digits of launch year)	16	16	18	18
12-14	International Designator (Launch number of the year)	083	083	002	002
15-17	International Designator (Piece of the launch)	A00	B00	A00	B00
19-20	Epoch Year (Last two digits of year)	21	21	21	21
21-32	Epoch (Day of the year and fractional portion of the day)	160.73	263.86	263.85	263.88
34-43	First Time Derivative of the Mean Motion	1.02E-05	1.21E-05	6.80E-06	8.71E-06
45-52	Second Time Derivative of Mean Motion (decimal point assumed)	000000-0	000000-0	000000-0	000000-0
54-61	BSTAR drag term (decimal point assumed)	056845-4	066665-4	039226-4	049311-4
63	Ephemeris type	0	0	0	0
65-68	Element number	0999	0999	0999	0999
69	Checksum	7	9	0	9
01	Line Number of Element Data	2	2	2	2
03-07	Satellite Number	41907	41908	43099	43100
09-16	Inclination	97.50	97.41	97.47	97.47
18-25	Right Ascension of the Ascending Node	349.16	334.40	342.09	342.30
27-33	Eccentricity	1.64E-03	1.54E-03	5.11E-04	1.10E-03
35-42	Argument of Perigee	82.01	78.17	335.09	320.80
44-51	Mean Anomaly	52.73	33.30	162.46	173.56
53-63	Mean Motion	15.16	15.16	15.16	15.16
64-68	Revolution number at epoch	26149	26149	20444	20445
69	Checksum	3	2	0	8

6.1.2 Operation and control system of "SuperView-1"

The commercial operation of remote sensing satellites is a significant trend in the operation and control of the space industry. It is estimated that "SuperView-1" gains more than 1,000 yuan per second of economic profits [183]. The goal of maximizing profit under the commercial operation model has given rise to in-depth research on "SuperView-1" task planning. In this section, the operation and control process, task planning models, and methods of "SuperView-1" are organized. The characteristics and limitations of the current systems, models, and algorithms are reviewed.

(1) Operation and control process analysis

The operation control process of the current "SuperView-1" commercial remote sensing constellation can be summarized as follows: the system maintains the task execution scheme of each satellite for a certain period of time in the future. It is always in a state to receive new user requirements. When the system function is activated, it determines the activation conditions—whether the system receives a new task or reaches the end of the previous planning cycle. If the system is activated because of the end of the previous planning cycle, the task and scenario parameters in the next cycle are updated according to the current information, and it is judged whether the scenarios in the next planning cycle are the same as the scenarios corresponding to the currently maintained task planning scheme. No operation is required if the scenario and task information have not changed. Otherwise, it goes directly to task pretreatment and subsequent processes. If the system is activated because of a new user requirement, it is necessary to determine whether the task is an emergency task and whether there is a measurement and control opportunity that can satisfy the task. When all the conditions are met, enter the task pretreatment and subsequent processes.

The basic operation control process of the "SuperView-1" constellation can be described in Figure 6.1. According to this figure, the operation and control process of the "SuperView-1" constellation can be more intuitively divided into two phases: the process before the task pretreatment is the data preparation phase before the task planning, and the process after the task pretreatment is the task planning phase. The main function of the data preparation stage is to ensure the validity of the data input into the task planning process through a series of judgment mechanisms to prevent falling into the invalid planning process to improve the overall operational efficiency of the system; the task planning stage forms the required task planning scheme according to the screened scenarios and tasks. In order to ensure the smooth operation of the whole process, users, satellite operation centers, satellite measurement and control centers, and other departments collaborate. After a long development period, this process has become a fixed mode for the satellite industry and application departments to deal with the corresponding business. Therefore, to realize the application goal in engineering projects, the characteristics and boundaries of the operation process must be considered when

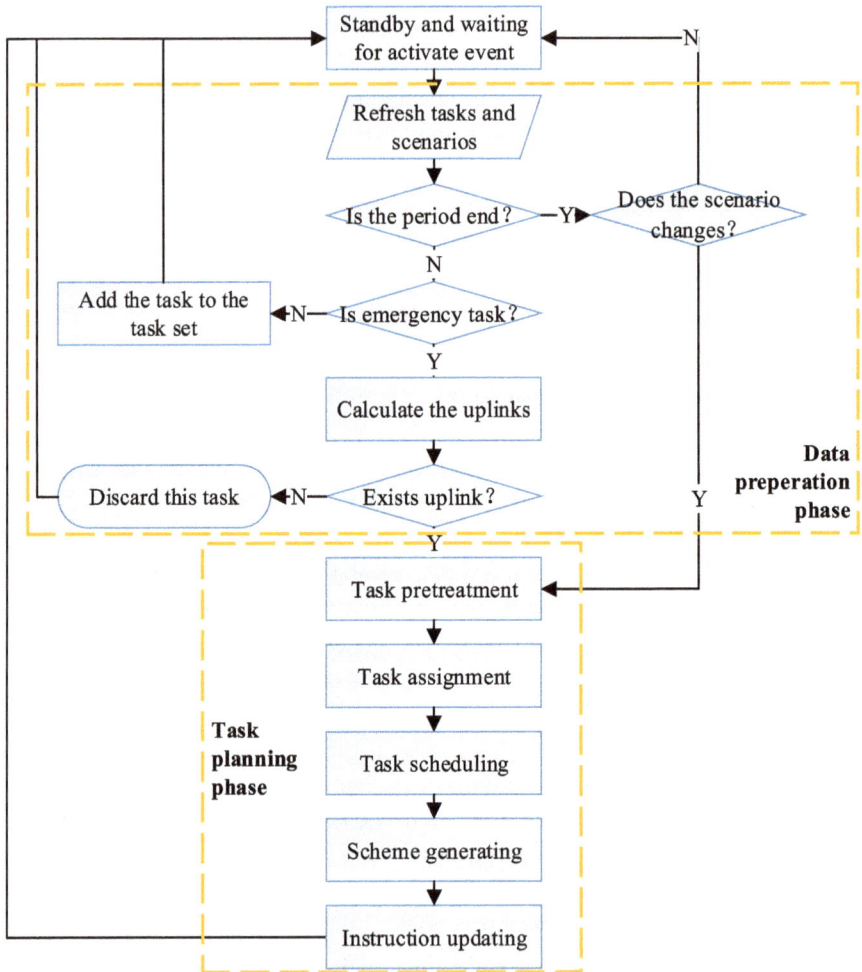

Figure 6.1: Operation control flowchart of "SuperView-1."

designing the task planning system. In order to guarantee the regular operation of the "SuperView-1" constellation, the collaboration process among the task planning system, measurement and control center, measurement and control station, data transmission station, and the constellation can be shown in Figure 6.2, which shows the process of the task planning system.

Figure 6.2 illustrates, with the help of a Gantt diagram, how the various parts of the whole system work as the timeline advances and the intrinsic relationships between the different activities among the parts. In the figure, the collaboration of the parts in the regular task planning scenario and the actions to be performed by each part are represented before the moment of arrival of the emergency task; when the task planning system receives the emergency task, it immediately activates the emergency task

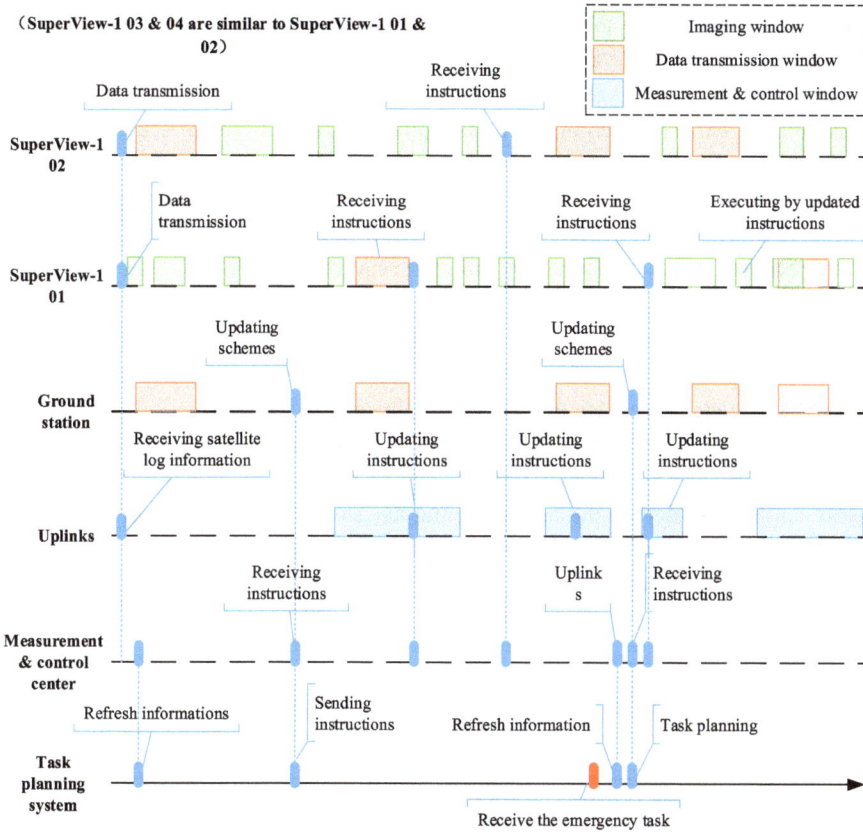

Figure 6.2: Collaboration of "SuperView-1" constellation operation.

planning process and requests the cooperation of various departments, such as the measurement and control center and the ground station, in order to complete the process of imaging the emergency task and the data transmission.

Through the analysis of the operation and control process of the "SuperView-1" constellation, it is found that the keys to the task planning system design can be summarized as follows:

① It is capable of managing all four satellites and their payloads and coordinating the external resources of the system, such as measurement and control stations and data transmission stations, which require unified management of all resources.

② It can support regular and emergency task planning procedures. In the regular procedure, the system receives user requirements in a batch and then plans tasks uniformly, which puts forward high requirements on the quality of the planning scheme; in the emergency procedure, the system receives emergency tasks and responds quickly, which puts forward high requirements on the time efficiency of the task planning algorithm.

③   Unified modeling and solving for different satellites and decoupling specific con-
    straints and constraint values from the optimization process as much as possible to
    reduce the cost of management and maintenance. To facilitate the subsequent real-
    ization of a generic system that can be applied to new satellite resources within a
    shorter development cycle and at a minor cost.
④   To deal with various types of user requirements, such as observation requirements
    for point targets, area targets, and stereo imaging. To carry out integrated analysis
    and comprehensive task planning for these requirements.

(2) Current planning model

The inputs, outputs, objective functions, and constraints of the current "SuperView-1"
task planning model have been organized as follows.

At present, the input and output parameters of the "SuperView-1" constellation for
task planning are also described as the task set and resource set, and the task planning
process can be regarded as the operation and decision-making process of the data in
the table. The task set includes a variable interface class and an attribute class, and the
resource set includes a platform class and a payload class, whose specific attributes are
shown in Table 6.2.

Table 6.2: Elements of the current SuperView 1 task planning model.

Composition	Class	Attributes
Task set	Variable interface class	Imaging start time, imaging end time, storage, data transmission start time, data transmission end time, etc.
	Attribute classes	Time required for imaging, time required for data transmission, record-play ratio, record rate, coordinate, profit, etc.
Resource set	Platform class	Satellite platforms, measurement and control stations, data transmission stations, etc.
	Payload class	Measurement antennas, data transmission antennas, imaging payloads, storage, battery, etc.

The constraints of the task planning model in the current system are organized
based on the satellite users' manual, which is refined in combination with the satellite's
hardware capability, control mode, functional positioning, and other conditions, taking into account the user's needs, by the satellite manufacturer, the satellite operation
center, and the measurement and control center. Constraints in practical engineering
problems are usually categorized by payload type, i. e., constraints satisfied by the same
payload are put together. It helps the satellite manufacturer and the operation center
understand the constraints. However, it is not conducive to discussing constraint classification, merger, and simplification. Therefore, the processing of constraints in engineering only translates the natural language into computer language without analyzing
and discussing the constraints in detail, which leads to a lot of repetitive computation

processes, which is one of the reasons affecting the operation efficiency of the task planning process in real-world engineering problems.

The objective function of the task planning model in the current system is to maximize the sum of the imaging profits of the "SuperView-1" constellation. It is determined by the profit-seeking nature of commercial remote sensing satellite operation, so the actual efficiency of the satellite must be maximized. Improving the task planning algorithm's computational efficiency and optimization effect is the core of improving the quality of the task planning scheme.

All satellites in the "SuperView-1" constellation are imaging satellites that have been developed and launched in recent years, and their payload capacity is at a high level, with high resolution, high revisit rate, high maneuver ability, high agility, and other advancements. It is because of these characteristics that make this problem model much more complex than the imaging satellite task planning model in the theoretical study. Its complexity is mainly reflected in the following three aspects:

① Decision-making actions are not limited to imaging tasks. In the actual satellite scheduling process, not only the decision-making of imaging tasks but also the operations such as data transmission and charging need to be considered. These actions also need to consider the relevant constraints. However, their impact on the state transfer is very different from that of the imaging task: the data transmission task consumes power but can release the storage space; the charging operation has no impact on the storage space and can increase the available power, but it is not permitted to perform any other action during charging.

② In the actual planning process, there are many more constraints to be considered than in the theoretical model, which are of various forms. There are dozens of constraints considered in the task planning process of "SuperView-1," and the constraints cover all the constraint types organized in Chapter 3, including both numerical comparison constraints and logical constraints. The large number of complex constraints brings significant challenges to the modeling process.

③ "SuperView-1 01" to "SuperView-1 04" are divided into two groups, and the constraints of these two groups differ. Therefore, in the modeling process, it is necessary to consider different constraints in a unified way to facilitate the planning algorithm.

(3) Current planning algorithm

Advanced satellite task scheduling algorithms can effectively improve the quality of satellites' work effectiveness so that the satellite can complete more tasks under limited resources. However, the complexity of the practical problems significantly reduces the efficiency of traditional operations research algorithms, so simple heuristic algorithms are usually applied to ensure their feasibility in practical engineering. The current "SuperView-1" task planning algorithm also divides the problem into task assignment and task scheduling.

The k-mean clustering algorithm is applied to the task assignment problem of "SuperView-1" constellation. The main idea of the algorithm is: first, read the complete task sequence; then randomly initialize four cluster centroids, then calculate the distance from each task to the cluster centroids, and assign the tasks to the cluster with the closest distance; after all the tasks are considered, recalculate the position of the cluster centroids based on the actual position of the tasks in each cluster, and repeat the above process until the cluster centroids converge. The pseudocode for this algorithm is as Algorithm 6.1.

Algorithm 6.1 K-mean clustering based task assignment algorithm.

Input: A set of tasks to be assigned
Output: Task assignment scheme
  1: Parameter initialization: $k = 4$;
  2: randomly generates k cluster centroids (i. e., one of the satellite's orbits);
  3: **repeat**
  4:   **for** $i = 1 : |TS|$ **do**
  5:     **for** $j = 1 : 4$ **do**
  6:       Calculate the clustering metrics of task i to clustering center j according to the equation (6.1);
  7:       Assigns task i to the cluster with the smallest clustering metric;
  8:     **end for**
  9:   **end for**
10:   **for** $j = 1 : 4$ **do**
11:     recalculates the center of the cluster, i. e., the center of mass of all the tasks in cluster j; the
12:   **end for**
13: **until** Positional Convergence

In this algorithm, $|TS|$ represents the number of tasks contained in the task set, and the number of clusters $k = 4$ is set according to the number of satellites. The process of calculating the metrics is related to many practical factors, such as the distance between the target and the subsatellite track, the average imaging quality, the range of the visible time window of the task on different resources, and the cloud occlusion rate. All of these factors can be selected as guidelines for task assignment, which can be specific to the operator's preferences. In the "SuperView-1" task planning project, the regular task planning process usually adopts the coverage of the task on different satellites as the clustering index to cluster the tasks, which can be calculated by the conceptual equation (6.1):

$$\text{Task coverage} = \frac{\text{Length of overlaps on visible time windows}}{\text{The length of visible time window}} \tag{6.1}$$

Algorithm 6.1 is a constructive heuristic algorithm where the clustering metrics are usually heuristic functions designed in conjunction with domain knowledge. The method has the distinct advantage that it is highly interpretable and allows for quick access to feasible schemes. The core part of the method is the design of the clustering criteria. However, the clustering criteria involved in this project are usually obtained based on personal experience or preference, and it is challenging to find methodologies that can be used to guide how to design and select the clustering criteria in real problems. Therefore, it is not easy to ensure the reasonableness of the method in various complex scenarios.

The task scheduling process currently used in "SuperView-1" task planning system is divided into an imaging task scheduling process and a data transmission task scheduling process. The method can be summarized as a hill-climbing algorithm based on the optimal imaging timing. The pseudo-code for this algorithm is Algorithm 6.2. In the algorithm, the climbing step length and the number of climbing steps are determined empirically in advance. Tasks are added to the scheduling scheme according to their optimal time. When a new task is added without violating the constraints, the task is added directly. Otherwise, the execution moment of the task is attempted to be adjusted based on the hill-climbing policy.

The algorithmic process can also be effectively decoupled from complex constraints, but the algorithm cannot guarantee the quality of the solution. On the other hand, the algorithm attempts a blind search process, resulting in low computational efficiency. The experimental section of this chapter will discuss the algorithm's application in real task planning scenarios in detail.

6.2 System design

"SuperView-1" task planning system is the core part of the constellation's operation system. It is also the "center" for satellites' daily control, operation, and maintenance. Starting from the actual operation control process of "SuperView-1" constellation and following the characteristics and customs of the current operation process, this subsection designs the external and internal interfaces of "SuperView-1" task planning system, as well as the data structure of the task planning, in order to provide a novel solution for planning tasks of "SuperView-1."

6.2.1 External interface design

The external interface design of "SuperView-1" task planning system is shown in Figure 6.3, and details of the interface content and usage are shown in Table 6.3. The primary interface relationships between "SuperView-1" task planning system and other parts of the operation control system (operation system, data transmission system) and

---

Algorithm 6.2 Hill climbing algorithm based on optimal imaging quality.

Input: A set of tasks to be scheduled

Output: Task scheduling scheme

1: Parameter initialization: climb step length s, climb step count c.

2: Sort tasks by the time window;

3: **while** task sequence is not empty **do**

4:     Calculates the optimal imaging time according to the equation (4.4);

5:     Attempts to determine the start time of the task based on the optimal imaging time;

6:     **if** The whole program does not satisfy the constraint after adding this task **then**

7:         **for** $k = 1 : c$ **do**

8:             Task start time = optimal imaging timing $+ k * s$;

9:             **if** Adjusted program does not satisfy constraints **then**

10:                 continue;

11:             **else**

12:                 Accept the task at the moment of the current attempt i and jump out of the current loop;

13:             **end if**

14:         **end for**

15:         **for** $k = 1 : c$ **do**

16:             task start time = optimal imaging timing $- k * s$;

17:             **if** Adjusted program does not satisfy constraints **then**

18:                 continue;

19:             **else**

20:                 Accept the task at the moment of the current attempt i and jump out of the current loop;

21:             **end if**

22:         **end for**

23:     **else**

24:         Accept the task i and update the scheduling program;

25:     **end if**

26: **end while**

---

external systems (control system) are shown in the figure. In order to facilitate understanding, the figure also summarizes the primary interface relationships between users, ground stations, the satellites, and their corresponding systems.

The task planning system, the operations system, and the receiving resources control system are the main components of the ground support system of "SuperView-1" constellation. The task planning system obtains imaging tasks from the operations system as one of the input parameters for planning. After receiving the imaging demands submitted by the satellite users, the operations system obtains the imaging tasks through

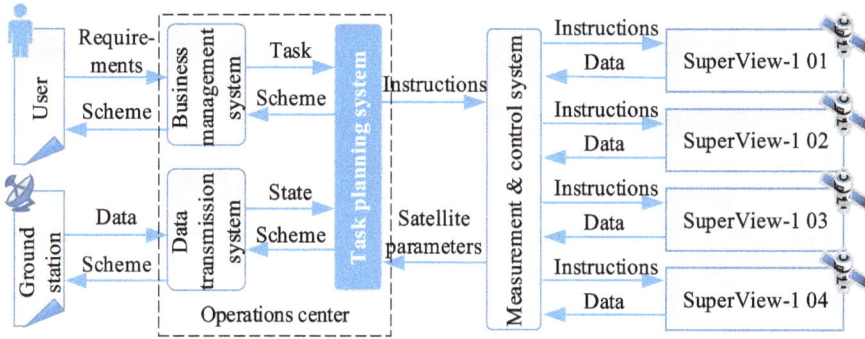

Figure 6.3: Schematic diagram of the external interface to the task planning system.

Table 6.3: Description of external interfaces to the task planning system.

Interface name	Sender	Receiver	Content and usage
Imaging tasks	Operations system	Task planning system	Task planning system receives imaging tasks standardized by the operations system, including requirements for imaging time, geographic location, clarity, imaging modes, work modes, and more
Planning scheme	Task planning system	Operations system	Task planning system sends a task planning scheme to operations system, including scheme information and specific parameter configuration information for each task
Receiving resources	Data transmission system	Task planning system	Data transmission system sends the receiving resources utilization to task planning system. The task planning process develops an appropriate data transmission plan based on this
Data transmission request	Task planning system	Data transmission system	Task planning system forms a data transmission plan, based on which a data transmission request needs to be sent to the data transmission system, requesting the corresponding data transmission station to cooperate with the process of data transmission
Programmed instructions	Task planning system	Satellite control system	Task planning system organizes a task planning scheme for each resource, which is then compiled into programmed instructions to be sent to measurement and control center
Satellite parameters	Satellite control system	Task planning system	Satellite control system corrects the real-time parameters of the resources based on the telemetry data fed back from the "SuperView-1," including the satellite's orbital parameters, payload parameters, and logs. Then sends them to the task planning system, which is one of the necessary information for task planning

a series of treatments, including demand management, demand acceptance and analysis, and task standardization. The task planning system obtains the receiving resource information from the operations system and generates the corresponding task planning scheme according to relevant information. The task planning system sends the scheme to the operations system to notify the satellite users of the planning results. On the other hand, it sends the application for the use of the data transmission resources in the scheme to the data transmission system to generate the data transmission scheme of each data transmission station.

The satellite control system is independent of the satellite operations system and bridges the ground support system and "SuperView-1." In this system, each task planning scheme is transformed into a control plan for each satellite, which is then compiled into programmed instructions that can be recognized by each "SuperView-1" satellite. The task planning system sends the programmed instructions to the satellite control system, which after reviewing and approving them, uploads the programmed instructions to the corresponding satellites through an uplink. The satellites execute the program control instructions by calling the platform and payload modules to work and then generate remote sensing data. The satellite control system receives the telemetry data from the satellites, updates, and organizes information on orbital elements and the state of task execution, and feeds it back to the task planning system to provide information support for the subsequent task planning process.

6.2.2 Internal interface design

Figure 6.4 gives the interface relationships between the main functional modules within the task planning system, as well as some of the external interfaces directly related to these functional modules. The sender, receiver, content, and usage of each internal interface are detailed in Table 6.4. Similar to the conventional remote sensing satellite

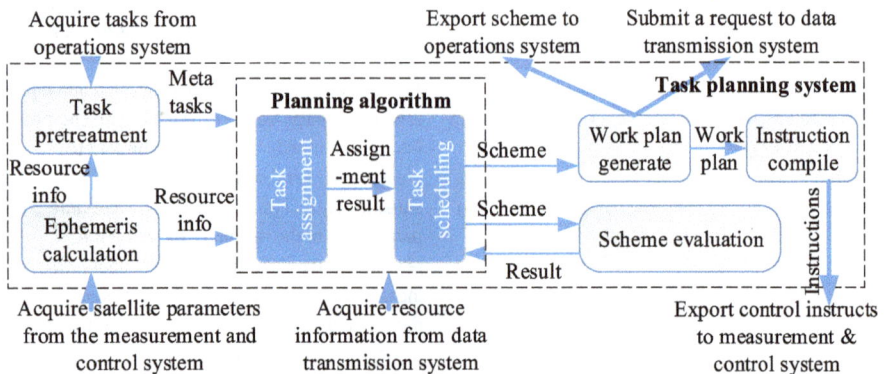

Figure 6.4: Schematic diagram of the internal interface of the task planning system.

Table 6.4: Description of the internal interfaces of the task scheduling system.

Interface name	Sender	Receiver	Content and usage
Metatask	Task pretreatment module	Task planning module	Task pretreatment module transforms the imaging task into a metatask and sends it to the task planning module, which provides information about the task's visible window of time, profit, and imaging duration of the task planning process
Planning scenarios	Task planning module	Task scheduling module, Performance evaluation module	Based on the received task and resource parameters, the task planning module generates a task planning scenario that satisfies the constraints and sends it to the performance evaluation module and the task scheduling module
Payload scheme	Task planning module	Instruction generation module	Task planning scheme is organized into the payload scheme of the satellite and the data transmission scheme of the station. These schemes are used to generate the payload scheme and send it to the instruction generation module. On the other hand, the task planning scheme and the data transmission scheme are sent to the operations system and the data transmission system, respectively
Forecast result	Orbital elements calculation and forecast module	Task pretreatment module, Task planning module	According to the orbit parameters provided by the control center, the orbital elements calculation and forecast module calculates the ephemeris and orbital information during the planning cycle, providing basic data support for the task pretreatment and task planning process
Evaluation results	Performance evaluation module	Task planning module	Performance evaluation module receives the planning scheme and evaluates the scheme based on the set metrics, then feeds the results back to the task scheduling module, which can be used to guide the improvement of the scheduling program
Assignment scheme	Task assignment module	Task scheduling module	Task assignment algorithm allocates tasks to appropriate imaging resources after receiving meta-task and resource information. The assignment result is one of the input conditions to the task scheduling algorithm

task planning system, "SuperView-1" task planning system consists of seven functional modules, including orbital elements calculation and forecast module, task pretreatment module, task assignment module, task scheduling module, task planning module, command generation module, and performance evaluation module. In addition to the above seven functional modules, there are other auxiliary functional modules such as task management module, resource management module, simulation and deduction module, etc., which are designed to improve the user experience. These functional modules do not directly affect the main process of task planning, so these will not be discussed in this book.

The task scheduling system consists of different function modules, in which the task planning module plays a central leading role in the system: the task planning module can directly call the data transmission information in the data transmission system. The task pretreatment module and the orbit calculation and forecasting module receive the imaging task and satellite parameters from the operation management system, respectively, and process this information into the metatask and satellite orbit forecasting results for the planning algorithm to call. The resource information managed by each system is transferred to the task planning module, such as data receiving resources, satellite measurement and control resources, and on-board payload resources, which are obtained through scientific calculation. Therefore, the computational efficiency and solution accuracy of the planning algorithm determine the operational efficiency of the whole application system.

6.2.3 Data structure design

In "SuperView-1" task planning system, the data structure depicts the way the system stores and organizes data. Based on Figure 6.4, the basic data structure of "SuperView-1" task planning system is designed as Figure 6.5.

Figure 6.5 organizes the operational objects and their relationships in the system by a class diagram in the UML model. In the diagram, each class contains three parts: class name, class attributes, and class operations. Among them, the user requirement class, coordinate class, imaging task class, transmission task class, task class, pointing angle class, attitude maneuver class, ground station class, imaging satellite class, and planning scheme class are entity classes that represent the data objects in the system. Among them, the planning scheme class records the detailed information of the planning results, which is the final output of the system; the other entity classes are designed according to the functional requirements of the system.

Orbital element prediction class, task pretreatment class, and task planning class are control classes that ensure logical relationships between entity classes by means of corresponding functional functions. The attributes of control classes can be empty, and each control class consists of several related functional operations. Among them, orbital forecast belongs to an underlying algorithm, which provides essential data support for

<<Angle>>
-Roll angle:Float
-Pitch angle:Float
-Yaw angle:Float
-Time:DateTime
-指向角计算函数

<<Maneuver>>
-Start time:DateTime
-Transition time:Int
-Duration:Int

<<Observe>>
-Imaging para:String
-Pointing angle:Angle
-Number of orbit:Int
-Earliest start:DateTime
-Latest end:DateTime
-Imaging start:DateTime
-Imaging end:DateTime
-Maneuver para:Maneuver
-Maneuver time calculation

<<Scheme>>
-Scheme ID:Int
-Satellite ID:String
-Export time:DateTime
-Number of tasks:Int
-Task details:Task

<<Orbit>>
-Ephemeris calculation
-Earth shadow calculation

<<Plan>>
-Task assignment
-Task scheduling

<<Pretreatment>>
-Task decomposition
-Task synthesis
-Visible time window

<<Download>>
-Working mode:String
-Ground station ID:Int
-Earliest tran:DateTime
-Latest trans:DateTime
-Trans begin time:DateTime
-Trans duration:Int
-File ID:String
-Storage calculation
-Transition time calculation

<<Task>>
-ID:Int
-Task type:Int
-Imaging attributes:Observe
-Data attribute:Download

<<Requirements>>
-Coordinates:Dot
-Country:String
-Type:String
-Profit:Int
-Imaging quality:Int
-Working mode:String
-Imaging mode:String

<<Station>>
-Coordinates:Dot
-Country:String
-number of antenna:Int
-Code rate:Float

<<Satellite>>
-Storage:float
-Recording ratio:float
-Compression ratio:float
-imaging width:int
-Orbital elements:string

<<Dot>>
-Longitude:Float
-Latitude:Float
-Altitude:Float

Figure 6.5: Design of internal data structures for task planning systems.

the task pretreatment process through functional operations such as orbital elements calculation and Earth shadow calculation; the task pretreatment transforms the user's requirements into corresponding imaging tasks and data transmission tasks through functional operations such as target decomposition, target synthesis, and visibility judgment. The operations of task planning consider the specific parameters of each resource and task and then formulate a satisfactory task planning scheme.

The relationship between the various classes in the system can be organized as follows: the system stores the imaging requirements put forward by the user in the user requirements class, including information on the target's geographic location, imaging profit, and clarity demand. One of the user requirements may contain multiple-point target imaging; the task pretreatment module reads the user requirement information and generates imaging tasks and data transmission tasks corresponding to a series of operations. Visible time windows of imaging tasks and data transmission tasks can be calculated by users' requirements. The task class typically contains a number of imaging tasks and data transmission tasks, or it may be empty. Each entity in the data transmission class represents a ground station with a uniquely determined geographic location; the satellite class records several hardware parameters of the satellite that are directly related to the constraints of task planning. The task planning class reads the necessary attributes in the tasks, receiving information about resources and imaging satellites to develop a reasonable task planning scheme. A planning scheme contains some imaging tasks and data transmission tasks, and the imaging tasks and data transmission tasks in the planning scheme usually appear in pairs.

6.3 Problem modeling and solving

Based on all the previous designs in this book, combined with the actual characteristics of the "SuperView-1" constellation, this section establishes a bilevel task planning model and its learning-based algorithms for the "SuperView-1" constellation.

6.3.1 Bilevel optimization model

(1) Input parameter

According to the design in Chapter 3, the input parameters of the "SuperView-1" task planning problem are also composed of two parts: resource information and task information. Assuming that satellite i in the system contains m_i orbiting cycles during a task planning period, then the total number of resources in the task planning model composed of the "SuperView-1" constellation $m = \sum_{i=1}^{4} m_i$. Resource-related information is portrayed by the attributes, i. e., $C_1, C_2, \dots$. The physical meanings and values of these attributes are usually closely related to the specifics of the resource's capability parameters and constraints. Combined with the research and design in this chapter, the resource attributes of "SuperView-1 01" to "SuperView-1 04" are organized as shown in Table 6.5.

It is worth mentioning that some of the parameters are intrinsic properties of the resources, such as $C1$, $C2$, and $C3$, which take values of constants, whereas $C4$ to $C9$ are time-dependent variables. These attributes are directly or indirectly used to determine the judgment of some constraints in the task planning process.

Table 6.5: Resource attribute description.

ID	Name	Data type	Description
C1	Write data rate	Const	Size of storage space occupied by imaging task per unit time
C2	Ratio of write to read	Const	Ratio of write data to read data
C3	Available solid-state memory	Float	Size of onboard solid-state memory
C4	Cumulative imaging time	Float	Sum of imaging hours for all imaging tasks in the planning period
C5	Cumulative number of imaging tasks	Int	Sum of imaging tasks number for all candidate tasks in the planning period
C6	Cumulative number of maneuvers	Int	Sum of attitude maneuvers number in the planning period
C7	Roll angle	Float	The value of the satellite's roll angle at a certain time
C8	Pitch angle	Float	The value of the satellite's pitch angle at a certain time
C9	Yaw angle	Float	The value of the satellite's yaw angle at a certain time

The basic properties of the task in this problem are designed according to equation (3.6), i. e., a task is described by time window, profit, and imaging duration together with some other necessary parameters. These necessary parameters include:

① Satellite number, orbit number, metatask number, whether in Earth shadow, time window start time, time window end time, imaging duration, pitch angle set, roll angle set, and yaw angle set. These are calculated by the task pretreatment process.
② The task number and task profit will be decided by the user and operations center and will be deposited in the user's demand order.
③ The ratio of write to read, write data rate, and imaging satellite payload are determined based on the actual capabilities of each "SuperView-1" satellite, combined with the characteristics of the task requirements, and these attributes have been determined when the user puts forward the requirements.
④ Ground station number and data transmission window number are the number referring to the corresponding ground station and data transmission task window. Some data transmission tasks have special requirements, and some can be flexibly configured according to the actual situation of resources.

(2) Output parameters
The final output of the model is the task planning schemes of the "SuperView-1 01" to "SuperView-1 04." Most parameters in the output scheme do not need to be determined by the optimization algorithm, and they can be obtained directly by calling input parameters, intermediate parameters, or numerical calculations. There are only two variables in the output data that require decision-making:

① Correspondence between tasks and resources: i. e., on which satellite each task is performed.

② Scheduling scheme for tasks: i. e., the specific moment of execution of each task to be executed on the corresponding resource.

(3) Objective function and constraints
A bilevel planning model for "SuperView-1" constellation is developed here with specific constraints:

$$\max \sum_{\{i|r_i>0\}} p_i^{r_i} \tag{6.2}$$

s. t.

$$G_k^1(r) \leq 0, \quad k = 1, 2, \ldots, g^1 \tag{6.3}$$

$$G_k^2(r) \leq 0, \quad k = 1, 2, \ldots, g^2 \tag{6.4}$$

$$G_k^3(r, es) \leq 0, \quad k = 1, 2, \ldots, g^3 \tag{6.5}$$

$$G_k^4(r, es) \leq 0, \quad k = 1, 2, \ldots, g^4 \tag{6.6}$$

$$r = (r_1, r_2, \ldots, r_n) \in \Omega_1 \tag{6.7}$$

$$es = (es_1, es_2, \ldots, es_n) \in \Omega_2 \tag{6.8}$$

The model follows the actual operations and control process standards of "Super-View-1" constellation, then transforms the actual engineering problems into a bilevel optimization model, which has substantial academic research value and engineering reference significance, which can be applied in other engineering projects with only a small amount of changes according to the specific background, and has potent portability.

As analyzed in Subsection 3.3.3 of this book, the objective function of this problem is to maximize the sum of the profits of the tasks in all the scheduling schemes, computed by equation (6.2). The sets of inequalities (6.3) to (6.6) represent, respectively, the cumulative constraints, rolling constraints, task attribute constraints, and task correlation constraints in "SuperView-1" task planning problem. The g^1, g^2, g^3, and g^4 represent the number of constraint entries of the corresponding class, respectively.

Among the four classes of constraints, the cumulative constraints G_k^1 and the rolling constraints G_k^2 constitute the set of upper-level constraints G_k^u in the bilevel scheduling model (Model (3.26)). The set of upper-level constraints can be statistically obtained based on the task assignment in the final task planning scheme, which is usually independent of the specific execution moment of the task; the task attribute constraints G_k^3 and the task correlation constraints G_k^4 are the lower-level constrains G_k^l in Model (3.26), which must be based on the specific moments of execution of the task and the related tasks in order to realize the determination process of the constraints. Therefore, according to the design in Chapter 3, the "SuperView-1" task planning problem can be modeled as a bilevel optimization problem: first, consider G_k^1 and G_k^2 to solve the task assignment problem of "SuperView-1" while combining all the constraints to solve the task scheduling problem under the condition that the task assignment scheme is determined.

6.3.2 Learning-based planning algorithm

In our method, "SuperView-1" constellation task planning problem is considered as a bilevel optimization problem, in which the upper-level optimization process solves the task assignment problem while the lower-level optimization process solves the task scheduling problem. Combined with the experimental results in Chapters 4 and 5, learning-based bilevel optimization algorithms by integrating deep Q-learning with HADRT or DPTS are applied to solve "SuperView-1" task planning problem, in order to explore the convergence, generalization, capability, and capability boundary of the designed algorithms in the process of practical engineering applications.

Based on the experimental results in Chapter 5, the basic algorithmic composition of the two algorithms selected for solving the "SuperView-1" task planning problem is shown in Table 6.6.

Table 6.6: Imaging satellite task planning algorithms by integrating deterministic algorithms and deep Q-learning.

No.	Task assignment process	Task scheduling process	Algorithm name
1	DQN	DPTS	DQN_DP
2	DQN	HADRT	DQN_CH

The basic framework of the learning-based optimization algorithm applied to the "SuperView-1" constellation is basically the same as the one designed in Figure 3.9, so it will not be repeated here. Based on the analysis of the "SuperView-1" task planning problem, combined with the design of the related algorithms in Chapter 4 and Chapter 5, the key points for realizing DQN_CH and DQN_DP in the "SuperView-1" task planning problem can be listed as follows:

① The constraint checking module used in algorithms deals with actual constraints, and the algorithmic flow of the constraint checking algorithm is a general process, which is the same as that of Algorithm 4.1.

② The process of DPTS and HADRT are the same Algorithm 4.3 and Algorithm 4.2.

③ The design of actions in DQN is the same as that in Subsection 5.2.2 of this book, with two classes of actions: "Select a task for the current resource" and "End adding a task to the current resource." It is worth mentioning that the action "select a task" in the experiments in Chapter 5 only includes the imaging task, while the experiments in this chapter consider the decision-making of the imaging task and the data transmission task to meet the real engineering applications.

④ The idea and steps for state updating in DQN are the same as those designed in Subsections 5.2.1 and 5.2.3 of this book. The key to updating the state is describing the environment in the MDP model. The environment is realized by the task scheduling algorithms DPTS and HADRT in conjunction with the specific problem characteristics. In the "SuperView-1" task planning problem, the state updating process of DQN is more complicated than the previous simulation experiments.

⑤ The formula for calculating the short-term reward in DQN is the same as equation (5.11). The profit value of each task directly affects the design of short-term reward, which in turn affects the training efficiency. In the "SuperView-1" constellation, the profit value of the received imaging task is generally decided by the user and operations center; the profit of the data transmission task is set to 50, which is much larger than the profit of imaging tasks because only the image data can be transmitted the benefits can be obtained in practical problems. Designing the task's profit value in this way avoids the problem of sparse return [184]. On the other hand, it also guides the algorithm in prioritizing the data transmission task to transmit the data as soon as possible to enhance the process's fast responsiveness and release more satellite storage space in time for subsequent imaging satellite tasks.

⑥ The value function and internal configuration information in DQN use the research results in Chapter 5. The topology of the value function is the same as Figure 5.6, and the configuration schemes of the internal activation function, loss function, and optimizer are designed based on the experimental conclusions of Subsection 5.2.5 of this book.

6.4 Simulation experiment

The algorithms designed in this chapter are implemented in Python, and all experimental procedures were conducted on a laptop with an Intel(R) Core(TM) i7-8750H CPU, 16.0 GB of RAM, and NVIDIA GeForce GTX 1060.

6.4.1 Experimental scene

(1) Scene data source
In the case of task planning for "SuperView-1", the input data consists mainly of imaging tasks, data transmission tasks, resources, and constraint-related attributes.

1) *Imaging tasks*

The imaging tasks in "SuperView-1" commercial remote sensing constellation operations system are calculated based on the imaging requirements received from different users. The various task classes, working modes, and imaging modes of the actual engineering problems introduced in the project background of this chapter are eventually transformed into standard imaging tasks. In this problem, at least 17 attributes are re-

Table 6.7: Imaging task attribute description.

Attributes	Data types	Data sources
Satellite No.	Int	Read from task pretreatment
Task No.	Int	Read from user's requirement
Orbiting cycle No.	Int	Calculated by task pretreatment
Task profit	Float	Read from user's requirement
Ratio of write to read	Float	Select from Satellite specifications
Write rate	Float	Select from Satellite specifications
Metatask No.	Int	Calculated by task pretreatment
Ground station No.	Int	Read from user requirements or decide by algorithm
Data transmission window No.	Int	Read from user requirements or decide by algorithm
Whether in Earth shadow	Bool	Calculated by task pretreatment
Use of imaging payload	String	Selection in satellite specifications
Time window start time	DateTime	Calculated by task pretreatment
Time window end time	DateTime	Calculated by task pretreatment
Imaging duration	Float	Calculated by task pretreatment
Pitch angle set	Float	Calculated by task pretreatment
Roll angle set	Float	Calculated by task pretreatment
Yaw angle set	Float	Calculated by task pretreatment

quired to describe an imaging task. The data types and sources of each basic attribute in an imaging task are organized as shown in Table 6.7.

2) *Data transmission tasks*

The data transmission module can process the available data transmission opportunities into data transmission tasks based on the geographic location of the data transmission station and satellite information. The data transmission task is mainly described by nine attributes: station number, satellite number, data transmission window number, orbiting cycle number, whether in Earth shadow, whether transmitted in real-time, window start time, window end time, and window duration. Table 6.8 shows samples of data transmission task attributes.

Table 6.8: Data transmission task attribute (sample).

Station number	Satellite number	Data transmission window number	Orbiting cycle	Whether in Earth shadow	Whether can be transmitted in real-time	Start time	End time	Window duration
1663	1	1	12636	0	0	42808	43312	504
1703	1	2	12637	0	0	48178	48750	572
1723	1	3	12639	0	0	59132	59563	431
...	...	...	...	...	...	...	...	...

In addition, the schemes for designing the resource and constraint-related properties in this problem have been detailed in Section 6.3.1 of this chapter, respectively, so we will not repeat them here. At this point, the preparation for the experiment is complete.

(2) Experimental design
All the data in this experiment are taken from the real data generated by the actual operations system of "SuperView-1" constellation [183]. Based on the task planning data set in the real system, this experimental scheme integrates the raw data into 14 task planning scenarios, and the basic features of the scenarios are summarized in Table 6.9.

Table 6.9: "SuperView-1" task planning scenarios.

Scenarios No.	Scenarios name	Satellite	Planning period	Number of tasks	Number of data transmission windows
1	S1_D1_1	"SuperView-1 01"	24 h	113	7
2	S1_D1_2	"SuperView-1 02"	24 h	148	7
3	S1_D1_3	"SuperView-1 03"	24 h	55	8
4	S1_D1_4	"SuperView-1 04"	24 h	137	7
5	S1_D2_1	"SuperView-1 01"	48 h	190	9
6	S1_D4_1	"SuperView-1 01"	96 h	397	28
7	S1_D7_1	"SuperView-1 01"	168 h	635	47
8	S4_D1_1	"SuperView-1" constellation	24 h	272	35
9	S4_D1_2	"SuperView-1" constellation	24 h	345	32
10	S4_D1_3	"SuperView-1" constellation	24 h	580	35
11	S4_D1_4	"SuperView-1" constellation	24 h	686	37
12	S4_D2_1	"SuperView-1" constellation	48 h	617	55
13	S4_D4_1	"SuperView-1" constellation	96 h	1266	56
14	S4_D7_1	"SuperView-1" constellation	168 h	2228	167

Among the 14 scenarios, scenarios numbered 1 to 7 are imaging satellite planning scenarios for single "SuperView-1" satellite, and scenarios numbered 8 to 14 are cooperative task planning scenarios for "SuperView-1" constellation. These task planning scenarios are designed in different task distributions and satellite constraints. Therefore, the group experiments can control the variables in the scenarios and study the performance trend of the algorithms with a single factor in the scenarios. The specific scheme of experiment organization is shown in Table 6.10.

In Table 6.10, experimental groups 1 and 2 discuss the performance of the algorithms in single-satellite scenarios, while experimental groups 3 and 4 discuss the performance of the algorithms in multisatellite scenarios. The detailed experimental implementation, results, and conclusions of the single-satellite and multisatellite scenarios are discussed in detail in Subsection 6.4.2 and Subsections 6.4.3 of this chapter, respectively.

Table 6.10: Design of the experimental organization.

Experimental group	Scenarios	Description	Purpose
1	1, 2, 3, 4	Scenarios of single-satellite task planning under different task sets	Testing the changing pattern of algorithm performance under different task sizes and task distributions
2	1, 5, 6, 7	Scenarios of single-star task planning under the same satellite with different planning periods	Testing the changing pattern of algorithm performance with task sizes and planning periods length
3	8, 9, 10 ,11	Multisatellites task planning scenarios under different task sets	Testing the changing patterns of the algorithm's performance under complex constraint scenarios with task sizes and distributions
4	11, 12, 13, 14	Multisatellites task planning scenarios under different planning periods	Testing the capability boundaries of the algorithms and discuss the changing patterns of algorithm performance with task sizes and planning periods length

(3) Algorithm parameters

In addition to applying the DQN_CH and DQN_DP algorithms designed in this book to solve "SuperView-1" task planning problem, two comparative algorithms are introduced in this experiment: the adaptive parallel mode evolution algorithm (APMA) [183] and the task planning algorithm based on K-mean clustering and hill-climbing algorithm (HC). Among them, the task planning algorithm based on K-mean clustering and hill-climbing algorithm is the solution algorithm used in the operations system of "SuperView-1," and the adaptive parallel modal evolution algorithm APMA [183] is a recently published work. Table 6.11 summarizes these algorithms' idea and parameters.

With the input data and scenario parameters, combined with the introduction of the experimental organization scheme and algorithms, studies related to "SuperView-1" task planning can be carried out.

6.4.2 Results of single satellite planning

The DQN_CH, DQN_DP, APMA, and HC algorithms are applied to scenarios 1 to 7 in Table 6.9. Each algorithm is repeated 10 times in each scenario, and the total return of the scenarios obtained in each run is recorded. The total profits of the scheme obtained is recorded for each run, and the statistics of the scalar mean and standard deviation of the total profits of the scheme are summarized in Table 6.12.

Based on the Table 6.12, the following conclusions can be drawn:

① Among all the algorithms, DQN_DP gets the highest mean of the total profit in all scenarios, the DQN_CH algorithm outperforms the algorithms APMA and HC in most

Table 6.11: Basic information and main parameters of the algorithms used in the experiment.

Serial No.	Name	Abbreviation	Implementation steps	Main parameters
1	Task planning algorithm based on deep Q-learning and constructive heuristics	DQN_CH	See Figure 5.8	See Section 5.3
2	Task planning algorithm based on deep Q-learning and dynamic programming	DQN_DP	See Figure 5.8	See Section 5.3
3	Adaptive Parallel Modal Evolutionary Algorithm	APMA	See Literature [183]	1. Population size: 100 2. Number of iterations: number of tasks * 500
4	Task planning algorithm based on K-means clustering and hill climbing	HC	Combining Algorithm 6.1 and Algorithm 6.2	1. Hill climbing steps: 5 seconds 2. Number of iterations: number of tasks * 500

Table 6.12: Total profits analysis of four algorithms in single satellite task planning scenarios.

Scenarios	DQN_DP		DQN_CH		APMA		HC	
	Mean	Standard deviation	Mean	Standard deviation	Mean	Standard deviation	Mean	Standard deviation
S1_D1_1	79	0	79	0	77	1.764	74.3	1.337
S1_D1_2	89	0	89	0	87.5	2.278	78.3	3.020
S1_D1_3	55	0	55	0	53.3	0.456	52.1	0.738
S1_D1_4	89	0	88	0	85.7	1.789	77.2	4.185
S1_D2_1	101	0	99	0	99.1	1.792	98.5	6.485
S1_D4_1	283	0	261	0	225	3.716	201.5	16.944
S1_D7_1	596	0	549	0	–	–	512.2	13.067

of the scenarios, and the algorithm APMA's mean total profit is slightly higher than that of DQN_CH only in the scenario S1_D2_1. Compared to APMA and HC, the advantage of the two integrated algorithms increases with the problem size. These prove the superiority of the learning-based bilevel algorithms in terms of solution accuracy.

② The standard deviation of the total profits of DQN_CH and DQN_DP is 0. Because the value function is no longer updated when the training process is finished. There are no random variables in the testing process of reinforcement learning, so the algorithms DQN_CH and DQN_DP formed by integrating deterministic algorithms and reinforcement learning are free of randomness, i. e., the experimental results of the algorithms do not change when they are repeated several times in each scenario, which ensures the stability of the algorithms.

③ Based on the hardware resources of this experiment (i7-8750H CPU, 16.0 GB RAM, GTX 1060), Algorithm APMA could not get the final result in scenario S1_D7_1 due to memory overflow. In contrast, the other algorithms can successfully run in all single-satellite scenarios. It illustrates that the algorithm APMA requires more computational resources in large-scale task planning scenarios. In other words, when the computational resources are limited, the size of the scenarios solvable by Algorithm APMA is smaller than the other algorithms.

In addition, the program running time of the learning-base bilevel task planning algorithms remained stable over 10 repetitions of the experiment. In contrast, both algorithm APMA and algorithm HC showed program running time outliers in some experimental scenarios: algorithm HC in scenario S1_D1_2 and scenario S1_D1_4, algorithm APMA in scenario S1_D2_1 all have records of operation times much larger than the normal calculation time. The records related to outliers in the running time of the program in the intercepted experiments are shown in Table 6.13.

Table 6.13: Outliers and related records on programming time.

Algorithm	Scenario	Running time of the program recorded by repeating 10 trials (s)								
HC	S1_D1_2	36.1	39.7	43.8	39.3	37.1	41.7	47.9	<u>88.9</u>	41.2	91.9
HC	S1_D1_4	36.8	32.8	34.2	27.6	34.0	36.5	35.2	33.3	<u>73.4</u>	33.2
APMA	S1_D2_1	68.3	<u>329.3</u>	67.5	74.9	69.2	69.2	69.6	70.5	69.9	69.6

The underlined data in Table 6.13 are the outliers. HC and APMA both have outliers in the repeated experiments due to the following reasons: both HC and APMA belong to heuristic algorithms constructed based on random search strategy, and parallel strategies are usually used to improve the efficiency of these algorithms. The learning-based bilevel algorithms DQN_DP and DQN_CH have stable operation time in repeated experiments in all test scenarios, which indicates the stability of these two algorithms in terms of operation time, which is conducive to improving the user experience in the practical application of the process.

By dividing all the results in single-satellite task planning scenarios into two groups, the performance of the algorithms in different task sets and different satellite constraints, as well as the variation of the planning period can be analyzed and obtained. The mean total profits of the individual algorithms in experimental group 1 and group 2 are shown intuitively in Figure 6.6.

Figure 6.6a shows the performance of the four algorithms in four scenarios with different task distributions. Among them, scenario S1_D1_3 represents the undersubscription scenario, while the others are considered as the oversubscribed scenarios. The results in the figure illustrate that both DQN_DP and DQN_CH are able to get better total profits than APMA and HC in dealing with scenarios with different task distributions,

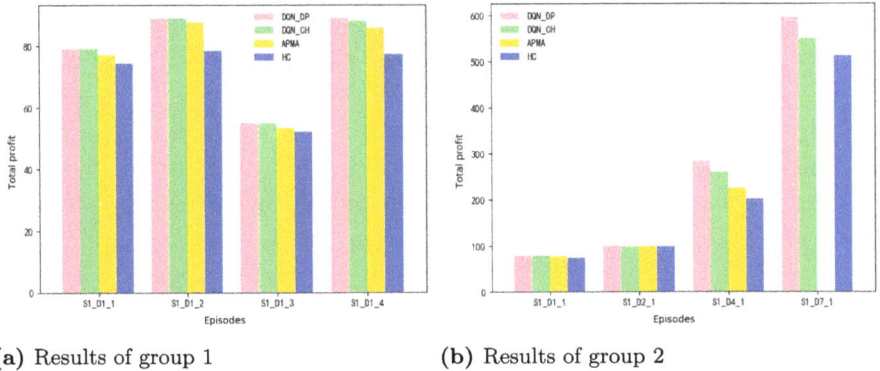

(a) Results of group 1   (b) Results of group 2

Figure 6.6: Experiment results of single satellite task planning scenarios.

which suggests that the algorithms have a high level of solution accuracy in different scenarios.

Figure 6.6b shows the performance of the four algorithms in scenarios with different planning periods and task sizes. As in the small-scale scenarios, all algorithms are able to obtain high-quality solutions. With the task size increasing, improving the solution results becomes harder. In scenarios S1_D1_1 and S1_D2_1, DQN_DP and DQN_CH still have an advantage in the solution accuracy by comparing to APMA and HC, indicating that the learning-based bilevel algorithms still have the advantage in this type of scenarios. The gap between the total profits of the DQN_DP, DQN_CH, and the other two algorithms gradually widens as the planning period increases. It is worth mentioning that in scenario S1_D7_1, the algorithm APMA cannot complete the computation because of memory overflow. It indicates that APMA needs more computational resources in scenarios with long planning period or large number of tasks, which is not conducive to generalizing its use in large-scale problems.

6.4.3 Results of multisatellite planning

DQN_CH, DQN_DP, APMA, and HC are applied to scenarios 8 to 14 in Table 6.9. Each algorithm is repeated 10 times in each scenario, and the total scheme returns obtained from each run are recorded. Each algorithm is repeated 10 times in each scenario, the total scheme profit obtained from each run is recorded, and the scalar mean and standard deviation statistics of the total profits are summarized in Table 6.14.

Based on Table 6.14, the following conclusions can be drawn:

① DQN_DP still has the best solution accuracy among all the algorithms in all multisatellite scenarios, and the total profits of algorithm DQN_CH in all scenarios outperforms that of APMA and HC. It proves that the two designed learning-based bilevel algorithms are still superior in terms of solution accuracy in multisatellite cooperative task planning scenarios.

Table 6.14: Total profits analysis of four algorithms in multisatellite cooperative planning scenarios.

Scenario	DQN_DP		DQN_CH		APMA		HC	
	Mean	Standard deviation	Mean	Standard deviation	Mean	Standard deviation	Mean	Standard deviation
S4_D1_1	241	0	239	0	236.1	2.132	222.1	3.635
S4_D1_2	263	0	259	0	258.6	3.239	236.3	6.255
S4_D1_3	299	0	298	0	295.3	3.683	280.8	3.910
S4_D1_4	342	0	337	0	330.7	4.596	301.6	7.516
S4_D2_1	551	0	538	0	–	–	468.5	7.106
S4_D4_1	1163	0	1116	0	–	–	–	–
S4_D7_1	–	–	2011	0	–	–	–	–

② In multisatellite scenarios, the standard deviation of the total profits of DQN_CH and DQN_DP is 0. It is an advantage of the deterministic algorithms, and the specific analysis has been given previously, so we will not repeat it here.

③ Based on the hardware resources in this experiment (i7-8750H CPU, 16.0 GB RAM, GTX 1060), all three algorithms, except DQN_CH, experienced memory overflow: APMA and HC could not get the final results in any of the planning scenarios with a planning period of more than 1 day and 2 days, respectively; DQN_DP cannot effectively obtain the results only in scenario S4_D7_1. Therefore, the learning-based bilevel algorithms require fewer computational resources than APMA and HC when solving large-scale problems to get satisfactory results.

The experimental results of the multisatellite collaborative task planning scenarios are visualized in Figure 6.7.

The testing scenarios in Figure 6.7a are multisatellite collaborative task planning scenarios with different task distributions and task sizes under the same planning period. Results show that the solution accuracy of HC is the lowest among all the scenarios, while that of the two learning-based bilevel algorithms is higher than the other two algorithms in all scenarios. Among them, the solution accuracy of DQN_DP is optimal in all scenarios.

Figure 6.7b tests the capability bounds of the algorithm. In the figure, DQN_CH finishes the solving process in all scenarios, and DQN_DP gets the result in the first three scenarios (S4_D1_1, S4_D2_1, S4_D4_1). HC solves the problem in scenario S4_D1_1 and scenario S4_D2_1 successfully, and APMA can only deal with scenarios S4_D1_1. By analyzing the performance of these algorithms in different scenarios, it can be found that after lengthening the planning period, the total profits of the tasks in the planning schemes obtained by all algorithms grows with it. Therefore, it can be judged that all the algorithms are able to effectively obtain satisfactory solutions in multisatellite collaborative task planning scenarios under different task planning periods, among which DQN_DP

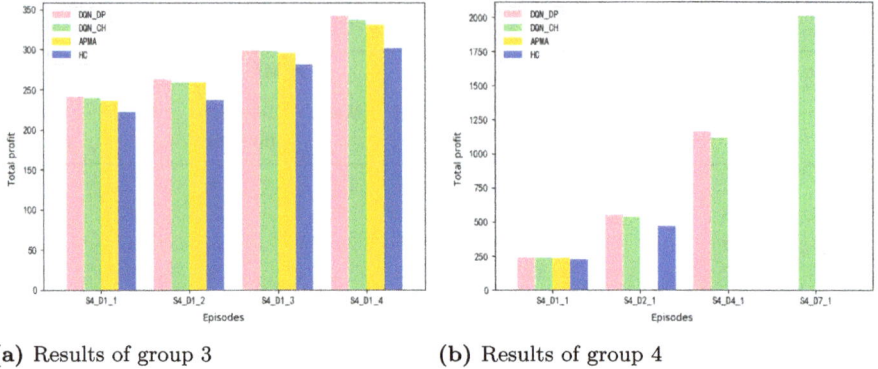

(a) Results of group 3   (b) Results of group 4

Figure 6.7: Experiment results of multisatellite task planning scenarios.

has the highest solution accuracy and DQN_CH has the lowest capacity requirement for computational resources.

Finally, the experiment also tests the load balance of each satellite in the solution scheme based on DQN_DP and DQN_CH. The experimental scenario is a satellite constellation composed of four "SuperView-1" satellites, and the satellite operations center always looks forward to the balance of tasks on each satellite so as to avoid large difference in the service life of the satellites in the constellation. Therefore, the payload balance of the satellites in the system is usually an evaluation index of concern in the practical application process.

The payload balancing of the scheme is discussed by analyzing the ratio of the number of tasks performed by each satellite during the planning period to the total number of tasks of the planning scheme, calculated according to the equation (6.9):

$$\begin{aligned} &\text{load balancing metrics of satellite } j \\ &= \frac{\text{number of tasks to be performed on satellite } j}{\text{total number of tasks in current program}} \end{aligned} \tag{6.9}$$

Table 6.15 shows the payload balancing metrics per satellite for different scenarios. All the data in the figure are between 23 % and 27 %, so the difference in the task number performed by each satellite is within 4 %. Since the four satellites in "SuperView-1" constellation are divided into two groups, the constraints of the satellites in the two groups are not the same, which leads to a slight difference in the number of tasks. Based on the experimental results, it can be concluded that the proposed algorithms can realize a satisfactory payload balance among satellites.

6.5 Summary

In this chapter, the Chinese's first commercial remote sensing constellation, "SuperView-1," is taken as the research object to verify the effectiveness of the bilevel op-

Table 6.15: Payload balancing analysis of satellites in multisatellite scenarios.

Scenarios	DQN_DP				DQN_CH			
	SuperView-1 01	SuperView-1 02	SuperView-1 03	SuperView-1 04	SuperView-1 01	SuperView-1 02	SuperView-1 03	SuperView-1 04
S4_D1_1	0.2407	0.2448	0.2614	0.2531	0.2469	0.2469	0.2469	0.2594
S4_D1_2	0.2433	0.2395	0.2586	0.2586	0.2432	0.2394	0.2587	0.2587
S4_D1_3	0.2408	0.2408	0.2609	0.2575	0.2450	0.2383	0.2617	0.2550
S4_D1_4	0.2456	0.2427	0.2544	0.2573	0.2374	0.2433	0.2611	0.2582
S4_D2_1	0.2450	0.2432	0.2559	0.2559	0.2416	0.2454	0.2546	0.2584
S4_D4_1	0.2442	0.2476	0.2519	0.2562	0.2455	0.2473	0.2518	0.2554
S4_D7_1	–	–	–	–	0.2481	0.2501	0.2511	0.2506

timization model and the learning-based task planning algorithms in practical engineering projects. The characteristics and difficulties of task planning in "SuperView-1" constellation are analyzed from the perspective of its operations and control systems, and then organizes the internal data flow, external data flow, and basic elements of the "SuperView-1" task planning system. Then, by combing the external and internal interfaces of the task planning system, a bilevel optimization model and its learning-based planning algorithms for "SuperView-1" are built.

The algorithm is compared with other commonly used algorithms in engineering or academic research, and the results of solution accuracy and payload balancing in real application scenarios proves that the application of integrating deterministic algorithms and reinforcement learning in real engineering can stably improve the average profit, and have more advantageous in the ability to deal with large-scale complex scenarios. The learning-based bilevel task planning algorithms have great advantages and development potential in solving imaging satellite task planning problems in real-world.

7 Summary and outlook

7.1 Summary

Accompanied by the deepening thinking of scholars in the field of aerospace applications and optimization on the imaging satellite task planning problem, the research on this problem gradually changes from orienting to theoretical models to serving the direction of practical applications. However, the increasing demand for multimode, multipayload satellite coordination planning in engineering has led to a further increase in the complexity of the imaging satellite task planning problems: the dimensionality of decision variables has been gradually increased, and the complexity and quantity of constraints have been increased, which leads to the challenge of exploring the analytical nature of the problem. On the other hand, the refinement of satellite management and control, as well as the user's expectation of rapid response to satellite imaging products put forward higher requirements on the process of imaging satellite task planning: Not only are there increasing expectations for the accuracy and efficiency of the task planning algorithm, but it can also be uniformly accessed and organically integrated with different types and natures of imaging satellites, thereby enhancing the algorithm's adaptability to different types of imaging satellites and improving its accuracy and efficiency. Based on the full investigation of existing satellite task planning technology, this book proposes a learning-based bi-level task planning theory and method to solve the imaging satellite task planning problem.

First, following the norms of computer software system design, the imaging satellite task planning system was designed. After defining the necessary terminology, starting from researching the typical foreign projects of imaging satellite task planning systems, combing the structure, functions, and interrelationships of the modules within the system of the typical foreign systems, thoroughly analyzing the design objectives and practical considerations of different projects, sorting out the system design in this study in terms of the overall application process, functional structure, business logic and other aspects of the needs, and then put forward the system design concept, determine the overall design idea and design principles; the unified modeling language (UML) is used to carry out the detailed design of the system's use case, structural objects and behavioral objects, which provides a methodology to managers and researchers in the satellite industry for understanding the satellite task planning process and designing the task planning system.

Second, we profoundly understand the essential characteristics of the imaging satellite task planning problem and then establish a bilevel optimization model to describe the problem: the upper-level task assignment process applies the value function trained by reinforcement learning to select the imaging opportunities of the task, while the lower-level task scheduling process uses deterministic algorithms to generate a stable and satisfactory task planning scheme. According to this idea, the complex imaging satellite task planning problem can be divided into two, which can effectively reduce the

https://doi.org/10.1515/9783111585109-007

overall difficulty of solving the problem, and ensure the quality of the solution and computational efficiency. The data generated by a large number of interactions between the upper and lower levels of the solving process can be used to continuously train the value function for task assigning: the assignment scheme output from the upper-level program is used as the input for the lower-level program, and the final scheme obtained from the task scheduling process can make up the data for training the value function. This framework is a new paradigm for solving satellite task planning problems, which can fully utilize the advantages of deterministic algorithms and reinforcement learning to improve the time efficiency of the algorithms while guaranteeing the quality of the solution.

Third, two deterministic algorithms are designed for the task scheduling problem. Specifically, based on the modeling and constraint analysis of the task scheduling process, a constraint checking method based on the timeline advancement mechanism is proposed and used as an underlying algorithmic module to support the operation of the task scheduling algorithms. The heuristic algorithm based on the density of residual task (HADRT) and the dynamic programming algorithm based on task sequencing (DPTS) is designed to solve the task scheduling problem. Both algorithms can obtain satisfactory solutions within polynomial time, which shows that the computational efficiency of these two algorithms is acceptable; the nature of deterministic algorithms ensures the stability of the algorithms' computational results; the treatment of constraints ensures the generality of the algorithms; and the optimality of the two algorithms under their specific assumptions is proved respectively. Simulation experiments prove the superiority of these two algorithms in terms of solution accuracy and efficiency. By comparing the differences between these two algorithms, the conclusions about applicable scenarios of each of the two algorithms are dug out: HADRT is more suitable for obtaining a satisfactory solution at a small cost of time in large-scale scenarios, whereas DPTS is suitable for further improving the quality of the solution in scenarios with more conflicts between tasks.

Fourth, a finite Markov decision process (MDP) model for the task assignment process is proposed. The solving process of task scheduling is nested in this model to ensure the integrity of the training process. By designing and analyzing the elements in the MDP model, an algorithmic solution framework oriented to random initial states and an action-pruning strategy based on domain knowledge are designed and applied to improve the Deep Q-Learning (DQN) algorithm. The ablation study of the algorithm is carried out to comprehensively investigate the impact of using different configurations in the DQN algorithm on the performance of the algorithm: first, the configuration scheme of the activation function, loss function, and optimizer in the improved DQN algorithm is determined, and the performance of the algorithm such as convergence and generalization are analyzed under this configuration scheme, so as to validate the effectiveness of the DQN in solving the task assignment problem. The performance of different algorithm integration schemes is investigated, and it is found that two task planning algorithms (DQN_DP and DQN_CH) outperform the other two classes of typical reinforcement learning algorithms (A3C and PtrN) in terms of convergence speed, convergence accuracy, and stability. Comparative analysis of DQN_DP and DQN_CH shows that DQN_CH has an advantage in computational efficiency, and DQN_DP is more advantageous in solution accuracy, so that the algorithms can be reasonably configured according to the actual needs to solve the actual complex applications.

This book verifies the effectiveness of the proposed models and methods in a real engineering problem, i. e., "SuperView-1" commercial remote sensing constellation task planning problem. The "SuperView-1" constellation task planning system conforming to the standards of operations and control process is designed, a bilevel optimization model for "SuperView-1" constellation task planning problem is established, and the learning-based bilevel planning algorithms by integrated deterministic algorithm and the reinforcement learning are proposed in this book. Case studies are carried out using real-world task planning scenarios, and the proposed algorithms are compared with two other algorithms that have been successfully applied in this project to analyze the application effects of the proposed algorithms in single-satellite task planning scenarios and multisatellite collaborative task planning scenarios, respectively. The results show that the proposed algorithms outperform two compared algorithms both in terms of solution accuracy and computational efficiency, and achieves satisfactory results in terms of payload balancing of the satellites, which demonstrates the rationality of the task assignment process. By analyzing the capability boundaries of the algorithms, it can be obtained that when the computational resources are limited, the integrated algorithm based on DQN and DPTS is more suitable for improving the solution quality in small-scale scenarios, while the integrated algorithm based on DQN and HADRT is more suitable for obtaining a satisfactory solution by using smaller computational resources in large-scale scenarios.

In conclusion, learning-based bilevel task planning models and algorithms can effectively alleviate the contradictions between versatility and efficiency, and solution efficiency and accuracy in imaging satellite task planning problems. They play an important role in narrowing the gap between theory and practice in the field of imaging satellite task planning.

7.2 Outlook

The content of this book is a preliminary exploration of the path of integrating machine learning and operations research algorithms to solve the complex imaging satellite task planning problem. After a series of studies, some exciting results and conclusions have been obtained, and it can be foreseen that, with the continuous maturity of the relevant underlying technology and scientific research conditions, this path is a promising research direction for a long time in the future. Facing the development trend of systematization, intelligence, refinement, and rapid response in the field of the task planning process of imaging satellites, future research work can be carried out at both theoretical and application levels.

In terms of theory, to further improve the solution accuracy and solution efficiency of the integrated algorithm:

(1) The deterministic algorithms designed in the current scheme are conditionally optimal because the imaging satellite scheduling problem mentioned in this book has been proven to be NP-hard, and the proposed deterministic approach cannot guarantee an optimal solution to the problem in polynomial time in any case. So, there is still room for improvement in the current algorithm, which can be further improved to guarantee optimality in more general situations.

(2) The impact of the decomposition approach in the bilevel optimization model on the whole solution process. According to the design of reference [71], the imaging satellite task planning problem can be further subdivided into three logically sequential solution processes: the matching problem of tasks and resources, the task sequencing problem, and the decision-making problem at the moment of task execution. However, the problem can be further divided into two solution stages by task types: imaging task scheduling and data transmission task scheduling. How does disassembling the problem in different ways affect the experimental results? Studying and answering this question is conducive to digging deeper into the inherent scientific laws of combining reinforcement learning algorithms with deterministic algorithms.

(3) Design schemes for improving the components in the MDP model. The MDP model designed in this book is mainly based on domain knowledge and research experience, but the model is still a preliminary model for the imaging satellite task planning problem. The structure and implementation methods of the elements in the model are relatively simple; for example, the elements in the algorithms such as the state representation method based on the original input parameters and the value function based on the three-layer fully-connected network, which are used in this scheme, can be studied in depth and further optimized. Whether other attributes can be used to describe the essential state of the scenario to improve the generalization ability of the model on the task scale, whether the convergence efficiency of the algorithm can be improved by designing a more complex short-term reward function and long-term value function are all scientific issues worthy of study.

In terms of application, to further expand the breadth of applications of bi-level optimization models and integration algorithms:

(1) Solving the large-scale imaging satellites and constellation task scheduling problem using bilevel optimization models and integrated algorithms [185, 186, 77]. Facing the large-scale imaging satellites and constellation task planning problem, the integrated algorithm designed in this book may have issues to solve, such as high computational cost and failure of convergence of the value function. Therefore, in order to adapt to the future development trend of large-scale imaging satellites and

constellation operations, the current bilevel optimization model and the integration algorithm need to be improved accordingly to cope with the problems of exploding solution space and low return rate of action selection during training.

(2) The problem of autonomous task planning for imaging satellites that take into account satellite dynamics [187, 25, 188, 189] is also one of the key directions of concern for researchers in the field of imaging satellite task planning. Since satellites encounter a variety of uncertainties during flight, imaging satellites can autonomously receive user requirements and make decisions accordingly when the hardware conditions and control means are mature. How the bilevel optimization model and the task planning algorithm integrating deterministic algorithm and reinforcement learning can make scientific decisions quickly under various uncertainties under such conditions is a topic worthy of in-depth research.

Description of symbols

Symbols of bilevel optimization models

i	Task ID
j	Resource ID
n	Number of tasks
m	Number of satellites
k	Number of orbiting cycles
TS	Task set
RS	Resource set
r_i	Resource ID to perform task i
C	Resource capacity parameter set
es_i	Execution start time of task i
ee_i	Execution end time of task i
ws_i^j	Start of visible time window for task i on resource j
we_i^j	End of visible time window for task i on resource j
d_i^j	Length of execution for task i on resource j
p_i^j	The profit of task i when executed on resource j
Ω	Solution space for the imaging satellite task planning problems
Ω_u	Solution Space for task assignment problems
Ω_l	Solution space for task scheduling problems
$G_{k_1}^u$	Set of constraints for the task assignment problem
$G_{k_2}^l$	Set of constraints for the task scheduling problem
g^u	Constraint number of the task assignment problem
g^l	Constraint number of the task scheduling problem

Symbols used in heuristic algorithms based on the density of residual tasks

TSQ	Complete Task Sequence
HQ	Current task sequence
$Stat_k$	kth cumulative constraints
$Roll_k$	kth rolling constraints
Att_k	kth task attribute constraints
$Corr_k$	kth task correlation constraint
L	Length of task scheduling period
k_s	Constraint number of cumulative constraints and rolling constraints
k_a	Constraint number of task attribute constraints and task correlation constraints
S_{opt}	Task planning scheme of the optimal solution
S_{HADRT}	Task planning scheme obtained by HADRT

https://doi.org/10.1515/9783111585109-008

Symbols used in dynamic programming algorithm based on task sequencing

x_i	State of stage i in a multi-stage decision model
X_i	Set of feasible states for stage i
$u_i(x_i)$	Decision variables of state x_i
$U_i(x_i)$	Set of feasible decisions for state x_i
$p_{i,j}(x_i)$	Strategy from stage i to stage j, under state x_i
$p_{i,j}^*(x_i)$	Optimal strategy from stage i to stage j, under state x_i

Symbols used in finite Markov decision models and improved deep Q-learning algorithms

t	Time step in the sequential decision-making process
N	Number of tasks
a_t	Action at time step t
S_t	State at time step t
R_t	Reward value at time step t
A	Action space
$A(S_t)$	Feasible set of actions under the state S_t
g^i	Geographical position information for task i
p^i	Profit of task i
v_t^i	Number of optional time windows remaining for task i after time step t
l_i^t	Availability metrics for task i after time step t

List of abbreviations

LEO	Low Earth Orbit
MEO	Medium Earth Orbit
HEO	High Earth Orbit
GEO	Geostationary Orbit
SAR	Synthetic Aperture Radar
AEOSSP	Agile Earth Observation Satellite Scheduling Problem
JSP	Job-Shop scheduling Problem
TSP	Traveling Salesman Problem
TSPTW	Traveling Salesman Problem with Time Window
VRP	Vehicle Routing Problem
RL	Reinforcement Learning
MDP	Finite Markov Decision Process (reinforcement learning)
UML	Unified Modeling Language (software engineering)
TLE	Two Line Elements
NASA	National Aeronautics and Space Administration (USA)
EO-1	Earth Observing1 (USA)
ASPEN	Automated Planning/Scheduling Environment (USA)
CASPER	Continuous Activity Scheduling Planning Execution and Replanning (USA)
DLR	Deutsches Zentrum für Luftund Raumfahrt, French Aerospace Center (France)
VAMOS	Verification of Autonomous Mission Planning Onboard a Spacecraft (France)
ESA	European Space Agency (EU)
PROBA	Project for On-Board Autonomy (EU)
CNES	Centre national d'études spatiales, French National Center for Space Studies (CNES)
AGATA	Autonomy Generic Architecture Test and Application (France)
FireBIRD	Fire BiSpectral Infrared Detection (France)
OBETTE	On-board Event Triggered Timeline Extension (France)
NP	Non-deterministic Polynomial (operations research)
CSP	Constraint Satisfaction Problem (operations research)
PDDL	Planning Domain Definition Language
ALNS	Adaptive Large Neighborhood Search (operations research)
HADRT	Heuristic Algorithm based on the Density of Residual Tasks (operations research)
HATW	Heuristic Algorithm based on Time-Window
DPTS	Dynamic Programming algorithm based on Task Sequencing
LACO	Learning-based Ant Colony Optimization
DQN	Deep Q-networks, Deep Q-learning algorithms (reinforcement learning)
A3C	Asynchronous Advantage Actor-Critic (Reinforcement Learning)
PtrN	Actor-Critic Algorithm with Pointer-networks (Reinforcement Learning)
SGD	Stochastic Gradient Descent (Deep Learning)
RMSprop	Root Mean Square Propagation (Deep Learning)
Adam	Adaptive Moment Estimation (Deep Learning)
MSE	Mean Squared Error (Deep Learning)
MAE	Mean Absolute Error (Deep Learning)
MAPE	Mean Absolute Percentage Error (Deep Learning)
MSLE	Mean Squared Log Error (Deep Learning)
HG	Hinge Loss (Deep Learning)
SH	Squared Hinge Loss (Deep Learning)
BC	Binary Cross-Entropy Loss (Deep Learning)

https://doi.org/10.1515/9783111585109-009

CC	Categorical Cross-Entropy Loss (Deep Learning)
KLD	Kullback Leibler Divergence (Deep Learning)
CP	Cosine Proximity (Deep Learning)

Bibliography

[1]   J. Shi et al. Space electronic technology space electronic technology space electronic technology. *Manned Spaceflight* 26 (4) (2020) 469–476 (in Chinese).

[2]   Y. He et al. A generic Markov decision process model and reinforcement learning method for scheduling agile Earth observation satellites. *IEEE Transactions on Systems, Man, and Cybernetics: Systems* 52 (3) (2022) 1463–1474.

[3]   H. Chen. Successful launch of SuperView-1 satellites. *Space Electronic Technology* 14 (3) (2017) 91 (in Chinese).

[4]   Z. You et al. Key technologies of smart optical payload in space remote sensing. *Spacecraft Recovery & Remote Sensing* 34 (1) (2013) 35–43 (in Chinese).

[5]   W. Lu. *Research on key technologies of space object situation cognition and service*. PhD thesis, Zhengzhou: Information Engineering University, 2020 (in Chinese).

[6]   H. Guo and L. Mao. SuperView-1 officially enters commercial use, bringing commercial remote sensing in China to the 0.5-metre era. *Satellite Application* 5 (2017) 62–63 (in Chinese).

[7]   S. Nag, A. S. Li and J. H. Merrick. Scheduling algorithms for rapid imaging using agile Cubesat constellations. *Advances in Space Research* 61 (3) (2018) 891–913.

[8]   F. Ip et al. Autonomous flood Sensorweb: Multi-sensor rapid response and early flood detection. In: 3rd Biennial Meeting of the International Environmental Modelling and Software Society: Summit on Environmental Modelling and Software, 2006.

[9]   G. Fasano and J. Pintér. *Modeling and Optimization in Space Engineering*. Springer, 2013.

[10]   G. Li. *Robust mission planning model and algorithm for multi-imaging autonomous satellites*. MA thesis, Harbin: Harbin Institute of Technology, 2018 (in Chinese).

[11]   X. Hu et al. A branch and price algorithm for EOS constellation imaging and downloading integrated scheduling problem. *Computers & Operations Research* 104 (2019) 74–89.

[12]   C. Verbeeck et al. A fast solution method for the time-dependent orienteering problem. *European Journal of Operational Research* 236 (2) (2014) 419–432.

[13]   Z. Zhang et al. A robustness task planning method for imaging satellite based on resources reservation. *System Engineering Theory and Practice* 36 (6) (2016) 1544–1554 (in Chinese).

[14]   X. Liu et al. An adaptive large neighborhood search metaheuristic for agile satellite scheduling with time-dependent transition time. *Computers & Operations Research* 86 (2017) 41–53.

[15]   Z. Lian et al. Integrating planning and scheduling for GNSS. In: *The 4th China Satellite Navigation Conference*, pp. 131–135. Wuhan, China, 2013 (in Chinese).

[16]   B. Ji, X. Yuan and Y. Yuan. A hybrid intelligent approach for co-scheduling of cascaded locks with multiple chambers. *IEEE Transactions on Cybernetics* 49 (4) (2019) 1236–1248.

[17]   R. Xu, X. Xu and P. Cui. Dynamic planning and scheduling algorithm based on temporal constraint network. *Computer-Integrated Manufacturing Systems* 10 (2) (2004) 188–194 (in Chinese).

[18]   H. Kona, A. Burde and D. R. Zanwar. A review of traveling salesman problem with time window constraint. *International Journal for Innovative Research in Science & Technology* 2 (1) (2015) 253–254.

[19]   B. Olli and M. Gendreau. Tabu search heuristics for the vehicle routing problem with time windows. *Top* 10 (2) (2002) 211–237.

[20]   B. Li et al. An overview and experimental study of learning-based optimization algorithms for vehicle routing problem. *IEEE/CAA Journal of Automatica Sinica* 9 (7) (2022) 1115–1138.

[21]   R. Daknama and E. Kraus. Vehicle routing with drones. arXiv:1705.06431, 2017.

[22]   S. Mitrovic-Minic et al. Collection planning and scheduling for multiple heterogeneous satellite missions: Survey, optimization problem, and mathematical programming formulation. In: *Modeling and Optimization in Space Engineering*, pp. 271–305. Springer, 2019.

[23]   N. Zufferey and M. Vasquez. A generalized consistent neighborhood search for satellite range scheduling problems. *RAIRO. Operations Research* 49 (1) (2015) 99–121.

https://doi.org/10.1515/9783111585109-010

[24]   J. Li. *Research on collaborative mission planning methods of autonomous multi-satellite: take geostationary orbit and low-orbit satellites collaborative mission as example*. PhD thesis, Changsha: National University of Defense Technology, 2017 (in Chinese).

[25]   J. Zhang and J. Dai. Simulation based autonomous satellite constellations design. In: *6th International Conference on System Simulation and Scientific Computing*, Beijing, China, pp. 1020–1024, 2005.

[26]   C. Chen. *Research on time-dependent scheduling methods and application on agile satellites mission planning*. PhD thesis, National University of Defense Technology, Changsha, 2014 (in Chinese).

[27]   Y. Du et al. Unified modeling and multi-strategy collaborative optimization for satellite task scheduling. *Control and Decision* 34 (9) (2019) 1847–1856 (in Chinese).

[28]   T. Benoist and B. Rottembourg. Upper bounds for revenue maximization in a satellite scheduling problem. *Quarterly Journal of the Belgian, French and Italian Operations Research Societies* 2 (3) (2004) 235–249.

[29]   L. Guo. *Research on key problems of agile satellite imaging scheduling problem*. PhD thesis, Wuhan: Wuhan University, 2015 (in Chinese).

[30]   W. Jiang, H. Hao and Y. Li. Review of task scheduling research for the Earth observing satellites. *Systems Engineering and Electronics* 35 (9) (2013) 1878–1885 (in Chinese).

[31]   J. Turner et al. Distributed task rescheduling with time constraints for the optimization of total task allocations in a multirobot system. *IEEE Transactions on Cybernetics* 48 (9) (2018) 2583–2597.

[32]   Y. Xu. *Joint scheduling for space-based maritime moving targets surveillance*. PhD thesis, Changsha: National University of Defense Technology, 2011 (in Chinese).

[33]   J. Wang, E. Demeulemeester and D. Qiu. A pure proactive scheduling algorithm for multiple Earth observation satellites under uncertainties of clouds. *Computers & Operations Research* 74 (2016) 1–13.

[34]   J. Wang et al. Exact and heuristic scheduling algorithms for multiple Earth observation satellites under uncertainties of clouds. *IEEE Systems Journal* 13 (3) (2019) 3556–3567.

[35]   J. Zhao et al. Granular prediction and dynamic scheduling based on adaptive dynamic programming for the blast furnace gas system. *IEEE Transactions on Cybernetics* 51 (4) (2021) 2201–2214.

[36]   L. Zhang. Research on the autonomous control architecture and planning and scheduling algorithms for satellite formation flying. *Computer CD Software and Applications* 8 (2010) 94 (in Chinese).

[37]   W. Zhang and T. G. Dietterich. High-performance job-shop scheduling with a time-delay $TD(\lambda)$ network. *Advances in Neural Information Processing Systems* 8 (1996) 1024–1030.

[38]   W.-J. Yin, M. Liu and C. Wu. A genetic learning approach with case-based memory for job-shop scheduling problems. In: *2002 International Conference on Machine Learning and Cybernetics*, pp. 1683–1687, vol. 3. IEEE, Beijing, China, 2002.

[39]   M. Nazari et al. Reinforcement learning for solving the vehicle routing problem. In: *Advances in Neural Information Processing Systems*, Montreal, Canada, pp. 9839–9849, 2018.

[40]   E.-G. Talbi. Machine learning into metaheuristics: a survey and taxonomy. *ACM Computing Surveys* 54 (6) (2021) 1–32.

[41]   M. Goerigk, A. Kasperski and P. Zieliński. Two-stage combinatorial optimization problems under risk. *Theoretical Computer Science* 804 (2020) 29–45.

[42]   J. N. D. Gupta, V. A. Strusevich and C. M. Zwaneveld. Two-stage no-wait scheduling models with setup and removal times separated. *Computers & Operations Research* 24 (11) (1997) 1025–1031.

[43]   Y. Tang, R. Liu and Q. Sun. Two-stage scheduling model for resource leveling of linear projects. *Journal of Construction Engineering and Management* 140 (7) (2014) 1–10.

[44]   A. K. Sadhu and A. Konar. Improving the speed of convergence of multi-agent q-learning for cooperative task-planning by a robot-team. *Robotics and Autonomous Systems* 92 (2017) 66–80.

[45]   M. A. Lopes Silva et al. A reinforcement learning-based multi-agent framework applied for solving routing and scheduling problems. *Expert Systems with Applications* 131 (2019) 148–171.

[46]   A. K. Sadhu and A. Konar. An efficient computing of correlated equilibrium for cooperative Q-learning-based multi-robot planning. *IEEE Transactions on Systems, Man, and Cybernetics: Systems* 50 (8) (2020) 2779–2794.

[47]   M. Leng. *Research on analysis methods of Earth-observing requests and application*. PhD thesis, Changsha: National University of Defense Technology, 2011 (in Chinese).

[48]   X. Zhang and S. Li. A multi-source remote sensing satellite demand modelling method for application-oriented topics. *Electronic Technology & Software Engineering* 6 (2016) 120–124 (in Chinese).

[49]   M. Ma et al. A model of Earth observing satellite application task describing. *Journal of National University of Defense Technology* 33 (2) (2011) 89–94 (in Chinese).

[50]   L. Shen, C.-J. Jiang and G.-J. Liu. Satellite objects extraction and classification based on similarity measure. *IEEE Transactions on Systems, Man, and Cybernetics: Systems* 46 (8) (2016) 1148–1154.

[51]   X. Liu et al. Task decomposition algorithm of remote sensing satellites based on space geometry model. *Systems Engineering and Electronics* 33 (8) (2011) 1783–1788 (in Chinese).

[52]   R. Sherwood et al. *Intelligent systems in space: The EO-1 autonomous sciencecraft*. Tech. rep. Arlington, United states: Jet Propulsion Laboratory, California Institute of Technology, 2005, pp. 150–160.

[53]   H. Reile, E. Lorenz and T. Terzibaschian. The FireBird mission – a scientific mission for Earth observation and hot spotdetection. In: *9th IAA Symposium on Small Satellites for Earth Observation*, Berlin, Germany, pp. 1–4, 2013.

[54]   M. A. Gleyzes, L. Perret and P. Kubik. Pleiades system architecture and main performances. In: *XXII ISPRS Congress*, Melbourne, Australia, pp. 537–542, 2012.

[55]   K. Luo and Y. Li. Planning overall high-resolution Earth observation system mission in the perspective of systems science. *System Engineering Theory and Practice* S1 (2011) 43–54 (in Chinese).

[56]   X. Ling. *Research on multi-satellite TT&C scheduling models and algorithms*. PhD thesis, Changsha: National University of Defense Technology, 2009 (in Chinese).

[57]   W.-C. Lin et al. Daily imaging scheduling of an Earth observation satellite. *IEEE Transactions on Systems, Man and Cybernetics. Part A. Systems and Humans* 35 (2) (2005) 213–223.

[58]   P. Wang, J. Li and Y. Tan. Design and implementation of a multi-satellite joint Earth observation capability assessment system. *Military Operations Research and Systems Engineering* 21 (2) (2007) 68–73 (in Chinese).

[59]   C. Pralet and G. Verfaillie. Using constraint networks on timelines to model and solve planning and scheduling problems. In: *18th International Conference on Automated Planning and Scheduling*, Sydney, Australia, pp. 272–279, 2008.

[60]   F. Xhafa et al. Genetic algorithms for satellite scheduling problems. *Mobile Information Systems* 8 (4) (2012) 351–377.

[61]   Y. Liu, Y. Chen and Y. Tan. The method of mission planning of the ground station of satellite based on dynamic programming. *Chinese Space Science and Technology* 25 (1) (2005) 44–47 (in Chinese).

[62]   Q. Ruan et al. Simulation-based scheduling for photo-reconnaissance satellite. In: *37th Winter Simulation Conference*, pp. 2585–2589. IEEE, Orlando, United States, 2005.

[63]   A. J. Vázquez Álvarez, R. S. Erwin et al. *An Introduction to Optimal Satellite Range Scheduling*. Springer Optimization and Its Applications vol. 106. Springer, 2015.

[64]   Z. Lian et al. Design of generic mission planning framework for high resolution Earth observation satellites. *Computer-Integrated Manufacturing Systems* 19 (5) (2013) 981–989 (in Chinese).

[65]   P. Tangpattanakul, N. Jozefowiez and P. Lopez. A multi-objective local search heuristic for scheduling Earth observations taken by an agile satellite. *European Journal of Operational Research* 245 (2) (2015) 542–554.

[66]   W. D. Ivancic et al. Secure, autonomous, intelligent controller for integrating distributed emergency response satellite operations. In: *2013 IEEE Aerospace Conference*, pp. 1–12. IEEE, Big Sky, United States, 2013.

[67]   D. Izzo and L. Pettazzi. Autonomous and distributed motion planning for satellite swarm. *Journal of Guidance, Control, and Dynamics* 30 (2) (2007) 449–459.

[68]   J. Barry. Increasing autonomy through satellite expert system scheduling. In: *AIAA 2nd Space Logistics Symposium*, pp. 153–156, 1988.

[69]   J. M. Barry and C. Sary. *Expert system for on-board satellite scheduling and control.* Tech. rep. NASA, Marshall Space Flight Center, Fourth Conference on Artificial Intelligence for Space Applications, 1988, pp. 193–203.

[70]   L. Barbulescu, A. Howe and D. Whitley. AFSCN scheduling: how the problem and solution have evolved. *Mathematical and Computer Modelling* 43 (9) (2006) 1023–1037.

[71]   W. J. Wolfe and S. E. Sorensen. Three scheduling algorithms applied to the Earth observing systems domain. *Management Science* 46 (1) (2000) 148–168.

[72]   X. Chen et al. A mixed integer linear programming model for multi-satellite scheduling. *European Journal of Operational Research* 275 (2) (2019) 694–707.

[73]   G. Peng et al. An exact algorithm for agile Earth observation satellite scheduling with time-dependent profits. *Computers & Operations Research* 120 (2020) 1–15.

[74]   Y. Bengio, A. Lodi and A. Prouvost. Machine learning for combinatorial optimization: a methodological tour d'horizon. *European Journal of Operational Research* 290 (2) (2021) 405–421.

[75]   G. W. Evans and R. Fairbairn. Selection and scheduling of advanced missions for NASA using 0-1 integer linear programming. *Journal of the Operational Research Society* 40 (11) (1989) 971–981.

[76]   S. Hartmann and D. Briskorn. A survey of variants and extensions of the resource-constrained project scheduling problem. *European Journal of Operational Research* 207 (1) (2010) 1–14.

[77]   J. Berger et al. A graph-based genetic algorithm to solve the virtual constellation multi-satellite collection scheduling problem. In: *2018 IEEE Congress on Evolutionary Computation (CEC)*, pp. 1–10. IEEE, Rio de Janeiro, Brazil, 2018.

[78]   G. Peng et al. Agile Earth observation satellite scheduling: an orienteering problem with time-dependent profits and travel times. *Computers & Operations Research* 111 (2019) 84–98.

[79]   D. Abramson. Constructing school timetables using simulated annealing: sequential and parallel algorithms. *Management Science* 37 (1) (1991) 98–113.

[80]   G. Bai. *Research on experimental satellite's mission planning and application for area targets survey.* PhD thesis, Changsha: National University of Defense Technology, 2009 (in Chinese).

[81]   J. Wang et al. Multi-objective imaging scheduling of Earth observing satellite based on constraint satisfaction. *Journal of National University of Defense Technology* 29 (4) (2007) 66–71 (in Chinese).

[82]   P. Baptiste et al. Constraint-based scheduling and planning. In: *Foundations of Artificial Intelligence*, vol. 2. Elsevier, 2006. Chap. 22.

[83]   A. Chen, Y. Jiang and X. Chai. Research on the formal representation of planning problem. *Computer Science* 35 (7) (2008) 105–110 (in Chinese).

[84]   Y. Liu. *Research on dynamic rescheduling model, algorithm and its applications of imaging reconnaissance satellite scheduling problem.* PhD thesis, Changsha: National University of Defense Technology, 2004 (in Chinese).

[85]   Y. Gao, L. Zhao and H. Yan. Modeling method of spacecraft system supported for autonomous mission planning and scheduling. *Journal of System Simulation* 21 (2) (2009) 320–324 (in Chinese).

[86]   S. Liu. *Autonomous planning for agile Earth-observing satellite integrating task and action.* PhD thesis, Changsha: National University of Defense Technology, 2017 (in Chinese).

[87]   S. Ackermann et al. Digital surface models for GNSS mission planning in critical environments. *Journal of Surveying Engineering* 140 (2) (2014) 1–11.

[88]   G. Beaumet, G. Verfaillie and M.-C. Charmeau. Feasibility of autonomous decision making on board an agile Earth-observing satellite. *Computational Intelligence* 27 (1) (2011) 123–139.

[89]   J. Li et al. Multi-objective evolutionary optimization for geostationary orbit satellite mission planning. *Journal of Systems Engineering and Electronics* 28 (5) (2017) 934–945.

[90]   W. Dong et al. A tissue P system based evolutionary algorithm for multi-objective VRPTW. *Swarm and Evolutionary Computation* 39 (2018) 310–322.

[91]   M. A. Adibi, M. Zandieh and M. Amiri. Multi-objective scheduling of dynamic job shop using variable neighborhood search. *Expert Systems with Applications* 37 (1) (2010) 282–287.

[92] Y. Chen. *Research on robust model and algorithm for mission planning of networking imaging satellite.* PhD thesis, Harbin: Harbin Institute of Technology, 2016 (in Chinese).

[93] G. Wu et al. Coordinated planning of heterogeneous Earth observation resources. *IEEE Transactions on Systems, Man, and Cybernetics: Systems* 46 (1) (2015) 109–125.

[94] S. Chien et al. Timeline-based space operations scheduling with external constraints. In: *20th International Conference on Automated Planning and Scheduling*, Toronto, Canada, pp. 34–41, 2010.

[95] L. Wang et al. Mission scheduling in space network with antenna dynamic setup times. *IEEE Transactions on Aerospace and Electronic Systems* 55 (1) (2019) 31–45.

[96] K. Sun et al. Adaptive fuzzy control for nontriangular structural stochastic switched nonlinear systems with full state constraints. *IEEE Transactions on Fuzzy Systems* 27 (8) (2019) 1587–1601.

[97] J. Qiu et al. Command filter-based adaptive NN control for MIMO nonlinear systems with full-state constraints and actuator hysteresis. *IEEE Transactions on Cybernetics* 50 (7) (2020) 2905–2915.

[98] L. N. Vicente and P. H. Calamai. Bilevel and multilevel programming: a bibliography review. *Journal of Global Optimization* 5 (3) (1994) 291–306.

[99] J. Lu et al. Multilevel decision-making: a survey. *Information Sciences* 346 (2016) 463–487.

[100] L. He et al. An improved adaptive large neighborhood search algorithm for multiple agile satellites scheduling. *Computers & Operations Research* 100 (2018) 12–25.

[101] X. Chu, Y. Chen and Y. Tan. An anytime branch and bound algorithm for agile Earth observation satellite onboard scheduling. *Advances in Space Research* 60 (9) (2017) 2077–2090.

[102] O. Vinyals, M. Fortunato and N. Jaitly. Pointer networks. In: *29th Annual Conference on Neural Information Processing Systems*, pp. 2692–2700. IEEE, Montreal, Canada, 2015.

[103] X. Chu. *Research on autonomous agile Earth observation satellite scheduling algorithm.* PhD thesis, Changsha: National University of Defense Technology, 2017 (in Chinese).

[104] P. Wang. *Research on branch-and-price based multi-satellite multi-station integrated scheduling method.* PhD thesis, Changsha: National University of Defense Technology, 2011 (in Chinese).

[105] B. Bai et al. Satellite orbit task merging problem and its dynamic programming algorithm. *Systems Engineering and Electronics* 31 (7) (2009) 1738–1742 (in Chinese).

[106] B. Dilkina and B. Havens. *Agile satellite scheduling via permutation search with constraint propagation.* Tech. rep. Vancouver, Canada: Actenum Corporation, 2005, pp. 1–20.

[107] Z. Xue. *Research on autonomous mission planning of Earth observation satellite.* MA thesis, Nanjing: Nanjing University of Aeronautics and Astronautics, 2015 (in Chinese).

[108] P. F. Maldague and A. Y. Ko. JIT planning: an approach to autonomous scheduling for space missions. In: *IEEE Aerospace Conference*, pp. 339–349. IEEE, Aspen, United States, 1999.

[109] S. Chien et al. Autonomous science on the EO-1 mission. In: *International Symposium on Artificial Intelligence Robotics and Automation in Space*, pp. 1–6. IEEE, Nara, Japan, 2003.

[110] Y. Chen et al. Bayesian optimization in AlphaGo. arXiv:1812.06855, 2018.

[111] K. Lachhwani and A. Dwivedi. Bi-level and multi-level programming problems: taxonomy of literature review and research issues. *Archives of Computational Methods in Engineering* 25 (4) (2018) 847–877.

[112] H. Qiu et al. Bi-level two-stage robust optimal scheduling for ac/dc hybrid multi-microgrids. *IEEE Transactions on Smart Grid* 9 (5) (2018) 5455–5466.

[113] B. Wille, M. T. Wörle and C. Lenzen. VAMOS—verification of autonomous mission planning on-board a spacecraft. In: *19th IFAC Symposium on Automatic Control in Aerospace*, Wurzburg, Germany, pp. 382–387, 2013.

[114] S. Chien et al. Planning operations of the Earth observing satellite EO-1: representing and reasoning with spacecraft operations constraints. In: *6th International Workshop on Planning and Scheduling in Space*, Darmstadt, Germany, pp. 1–8, 2009.

[115] B. Cichy et al. Validating the autonomous EO-1 science agent. In: *15th International Conference on Automated Planning and Scheduling*, Monterey, United States, pp. 39–47, 2005.

[116] S. Cheng and R. Gong. The first "Pleiades" satellite is scheduled to be launched this year. *Space International* 7 (2011) 11–16 (in Chinese).

[117] K. A. M. Goetz, F. Huber and M. von Schoenermark. VIMOS—autonomous image analysis on board of BIROS. In: *19th IFAC Symposium on Automatic Control in Aerospace*, Wurzburg, Germany, pp. 423–428, 2013.

[118] J. Zender et al. The Projects for Onboard Autonomy (PROBA2) Science Centre: Sun Watcher using APS detectors and image Processing (SWAP) and Large-Yield Radiometer (LYRA) science operations and data products. *Solar Physics* 286 (1) (2013) 93–110.

[119] H. Gao et al. Development of overseas Earth-observing satellite technology. *Spacecraft Engineering* 18 (3) (2009) 84–92 (in Chinese).

[120] S. A. Chien et al. Using iterative repair to improve the responsiveness of planning and scheduling. In: *5th International Conference on Artificial Intelligence Planning Systems*, Breckenridge, United States, pp. 300–307, 2000.

[121] H. Wang, H. He and Z. Yang. Scheduling of agile satellites based on an improved quantum genetic algorithm. *Acta Astronautica* 39 (11) (2018) 1266–1274 (in Chinese).

[122] Y. Chen et al. A learnable ant colony optimization to the mission planning of multiple satellites. *System Engineering Theory and Practice* 33 (3) (2013) 791–801 (in Chinese).

[123] A. Globus et al. Scheduling Earth observing satellites with evolutionary algorithms. In: *International Conference on Space Mission Challenges for Information Technology*, pp. 1–7, 2003.

[124] A. Globus et al. A comparison of techniques for scheduling Earth observing satellites. In: *19th National Conference on Artificial Intelligence*, San Jose, United States, pp. 836–843, 2004.

[125] D. Habet, L. Paris and C. Terrioux. A tree decomposition based approach to solve structured SAT instances. In: *21st IEEE International Conference on Tools with Artificial Intelligence*, pp. 115–122. IEEE, Newark, United States, 2009.

[126] X. Zhai et al. Robust satellite scheduling approach for dynamic emergency tasks. *Mathematical Problems in Engineering* 2015 (2015) 1–20.

[127] M. Wang, G. Dai and M. Vasile. Heuristic scheduling algorithm oriented dynamic tasks for imaging satellites. *Mathematical Problems in Engineering* 2014 (5) (2014) 1–11.

[128] J. Wang et al. Towards dynamic real-time scheduling for multiple Earth observation satellites. *Journal of Computer and System Sciences* 81 (1) (2015) 110–124.

[129] Z. Yan, Y. Chen and L. Xing. Agile satellite scheduling based on improved ant colony algorithm. *System Engineering Theory and Practice* 34 (3) (2014) 793–801 (in Chinese).

[130] F. Yao and L. Xing. Learnable ant colony optimization algorithm for solving satellite ground station scheduling problems. *Systems Engineering and Electronics* 34 (11) (2012) 2270–2274 (in Chinese).

[131] L. Xing and Y. Chen. Mission planning of satellite ground station system based on the hybrid ant colony optimization. *Acta Automatica Sinica* 34 (4) (2008) 414–418 (in Chinese).

[132] L. Xing and Y. Chen. Research progress on intelligent optimization guidance approaches using knowledge. *Acta Automatica Sinica* 37 (11) (2011) 1285–1289 (in Chinese).

[133] X. Chu, Y. Chen and L. Xing. A branch and bound algorithm for agile Earth observation satellite scheduling. In: *Discrete Dynamics in Nature and Society 2017*, 2017.

[134] Z. Ma. *Handbook of Modern Applied Mathematics: Volume on Operations Research and Optimization Theory*, vol. 1. Tsinghua University Press, 1998 (in Chinese).

[135] P. Shen. *Global Optimization Method*. Science Press, 2006 (in Chinese).

[136] C.-C. Tsai and S. H. A. Li. A two-stage modeling with genetic algorithms for the nurse scheduling problem. *Expert Systems with Applications* 36 (5) (2009) 9506–9512.

[137] A. Parisio and C. N. Jones. A two-stage stochastic programming approach to employee scheduling in retail outlets with uncertain demand. *Omega* 53 (2015) 97–103.

[138] F. Wu and R. Sioshansi. A two-stage stochastic optimization model for scheduling electric vehicle charging loads to relieve distribution-system constraints. *Transportation Research. Part B: Methodological* 102 (2017) 55–82.

[139] H. Dashti et al. Weekly two-stage robust generation scheduling for hydrothermal power systems. *IEEE Transactions on Power Systems* 31 (6) (2016) 4554–4564.

[140] L. He et al. Hierarchical scheduling for real-time agile satellite task scheduling in a dynamic environment. *Advances in Space Research* 63 (2) (2019) 897–912.

[141] B. Deng et al. Two-phase task scheduling in data relay satellite systems. *IEEE Transactions on Vehicular Technology* 67 (2) (2017) 1782–1793.

[142] G. Li. *Models and algorithms for distributed earth observation satellite system online coordination scheduling under communication constraints*. PhD thesis, Changsha: National University of Defense Technology, 2017 (in Chinese).

[143] C. Wang. *Distributed cooperative task planning research of Earth observing satellites based on agent*. PhD thesis, Changsha: National University of Defense Technology, 2014 (in Chinese).

[144] Z. Yu. *Research on key technologies for multiple satellite conters cooperative planning platform*. PhD thesis, Changsha: National University of Defense Technology, 2010 (in Chinese).

[145] Y. Miao. *Research on autonomous task planning of imaging satellite of formation flying*. PhD thesis, Harbin: Harbin Institute of Technology, 2016 (in Chinese).

[146] R. Grasset-Bourdel, G. Verfaillie and A. Flipo. Planning and replanning for a constellation of agile Earth observation satellites. In: *21th International Conference on Automated Planning and Scheduling*, Freiburg, Germany, pp. 29–36, 2011.

[147] S. Desale et al. Heuristic and meta-heuristic algorithms and their relevance to the real world: a survey. *International Journal of Computer Engineering in Research Trends* 2 (5) (2015) 269–304.

[148] G. Wu. *Research on the exploration and exploitation strategies controled intelligent optimization approaches and its application*. PhD thesis, Changsha: National University of Defense Technology, 2014 (in Chinese).

[149] Z. Zhou. *Machine Learning*. Tsinghua University Press, 2016 (in Chinese).

[150] H. Song, I. Triguero and E. Özcan. A review on the self and dual interactions between machine learning and optimisation. *Progress in Artificial Intelligence* 8 (2) (2019) 143–165.

[151] R. S. Sutton and A. G. Barto. *Reinforcement Learning: An Introduction*. MIT Press, 2018.

[152] D. Silver et al. Mastering the game of Go with deep neural networks and tree search. *Nature* 529 (7587) (2016) 484–489.

[153] D. Silver et al. Mastering the game of Go without human knowledge. *Nature* 550 (7676) (2017) 354–359.

[154] D. Silver et al. A general reinforcement learning algorithm that masters chess, shogi, and Go through self-play. *Science* 362 (6419) (2018) 1140–1144.

[155] M. Moravčík et al. Deepstack: expert-level artificial intelligence in heads-up no-limit poker. *Science* 356 (6337) (2017) 508–513.

[156] Z. Xu et al. Large-scale order dispatch in on-demand ride-hailing platforms: a learning and planning approach. In: *24th ACM SIGKDD International Conference on Knowledge Discovery and Data Mining*, London, United Kingdom, pp. 905–913, 2018.

[157] Z. Qin et al. Ride-hailing order dispatching at DiDi via reinforcement learning. *INFORMS Journal on Applied Analytics* 50 (5) (2020) 272–286.

[158] H. Hu et al. Solving a new 3d bin packing problem with deep reinforcement learning method. arXiv:1708.05930, 2017, pp. 1–7.

[159] M. Pinedo. *Scheduling: Theory, Algorithms, and Systems*. Springer, 2012.

[160] R. Grunitzki, G. de Oliveira Ramos, A. Lucia and C. Bazzan. Individual versus difference rewards on reinforcement learning for route choice. In: *2014 Brazilian Conference on Intelligent Systems*, pp. 253–258. IEEE, Sao Carlos, Brazil, 2014.

[161] Y. Li. Deep reinforcement learning: an overview. arXiv:1701.07274, 2017, pp. 1–85.

[162] S. Adam, L. Busoniu and R. Babuska. Experience replay for real-time reinforcement learning control. *IEEE Transactions on Systems, Man and Cybernetics. Part C, Applications and Reviews* 42 (2) (2012) 201–212.

[163] R. Cui et al. Adaptive neural network control of AUVs with control input nonlinearities using reinforcement learning. *IEEE Transactions on Systems, Man, and Cybernetics: Systems* 47 (6) (2017) 1019–1029.

[164] R. Sharma. Fuzzy Q learning based UAV autopilot. In: *The International Conference on Innovative Applications of Computational Intelligence on Power, Energy and Controls with Their Impact on Humanity*, pp. 29–33. IEEE, Ghaziabad, India, 2014.

[165] Y. Wang et al. Joint deployment and task scheduling optimization for large-scale mobile users in multi-UAV-enabled mobile edge computing. *IEEE Transactions on Cybernetics* 50 (9) (2020) 3984–3997.

[166] L. M. Gambardella and M. Dorigo. Ant-Q: a reinforcement learning approach to the traveling salesman problem. In: *12th International Conference on Machine Learning*, Tahoe City, United States, pp. 252–260, 1995.

[167] Y. Wei and M. Zhao. A reinforcement learning-based approach to dynamic job-shop scheduling. *Acta Automatica Sinica* 31 (5) (2005) 765–771.

[168] H. Dai et al. Learning combinatorial optimization algorithms over graphs. In: *Advances in Neural Information Processing Systems*, Long Beach, United States, pp. 6349–6359, 2017.

[169] K. Li, T. Zhang and R. Wang. Deep reinforcement learning for multiobjective optimization. *IEEE Transactions on Cybernetics* 51 (6) (2021) 3103–3114.

[170] H. Lu, X. Zhang and S. Yang. A learning-based iterative method for solving vehicle routing problems. In: *International Conference on Learning Representations*, Addis Ababa, Ethiopia, pp. 1–15, 2020.

[171] W. Usaha and J. A. Barria. Reinforcement learning for resource allocation in LEO satellite networks. *IEEE Transactions on Systems, Man and Cybernetics. Part B. Cybernetics* 37 (3) (2007) 515–527.

[172] C. Wang et al. An algorithm of cooperative multiple satellites mission planning based on multi-agent reinforcement learning. *Journal of National University of Defense Technology* 33 (1) (2011) 53–58 (in Chinese).

[173] W. Haijiao. *Massive scheduling method under online situation for satellites based on reinforcement learning*. PhD thesis, Beijing: University of Chinese Academy of Sciences (National Space Science Center), 2018 (in Chinese).

[174] H. Wang et al. Online scheduling of image satellites based on neural networks and deep reinforcement learning. *Chinese Journal of Aeronautics* 32 (4) (2019) 1011–1019.

[175] T. T. Nguyen, N. D. Nguyen and S. Nahavandi. Deep reinforcement learning for multiagent systems: a review of challenges, solutions, and applications. *IEEE Transactions on Cybernetics* 50 (9) (2020) 3826–3839.

[176] D. P. Kingma and J. Lei Ba. Adam: a method for stochastic optimization. In: *3rd International Conference on Learning Representations*, San Diego, United States, pp. 1–15, 2015.

[177] T. Schaul et al. Prioritized experience replay. In: *4th International Conference on Learning Representations*, San Juan, Puerto Rico, pp. 1–21, 2016.

[178] Y. He et al. Auto mission planning system design for imaging satellites and its applications in environmental field. *Polish Maritime Research* 23 (s1) (2016) 59–70.

[179] I. Bello et al. Neural combinatorial optimization with reinforcement learning. arXiv:1611.09940, 2016, pp. 1–15.

[180] Z. Liu. "SuperView-1" satellites have ushered in a new era of commercial remote sensing in China. *Spacecraft Recovery & Remote Sensing* 38 (2) (2017) 2 (in Chinese).

[181] C. Yang et al. "SuperView-1" is ready to go: China will enter the 0.5-metre commercial remote sensing market. *Installation* 2 (2017) 26 (in Chinese).

[182] E. Cui. Successful launch of "SuperView-1" China's first small satellite for secondary school students was launched on board. *Aerospace China* 1 (2017) 22 (in Chinese).

[183] Y. Du. *Research on the general-purpose scheduling engine for satellite task scheduling problems*. PhD thesis, Changsha: National University of Defense Technology, 2021 (in Chinese).

[184] M. J. Mataric. Reward functions for accelerated learning. In: *11th International Conference on Machine Learning*, pp. 181–189. Elsevier, New Brunswick, United States, 1994.

[185] H. Kim and Y. K. Chang. Mission scheduling optimization of SAR satellite constellation for minimizing system response time. *Aerospace Science and Technology* 40 (2015) 17–32.

[186] P. Wang et al. A model, a heuristic and a decision support system to solve the scheduling problem of an Earth observing satellite constellation. *Computers & Industrial Engineering* 61 (2) (2011) 322–335.

[187] J. Zhang and J. Dai. A hybrid agent oriented modeling method of autonomous Distributed Satellite Systems (DSS). In: *6th International Conference on System Simulation and Scientific Computing*, Beijing, China, pp. 4–7, 2005.

[188] D. Thomas et al. Real-time on-board estimation & optimal control of autonomous micro-satellite proximity operations. In: *55th AIAA Aerospace Sciences Meeting*, Grapevine, United States, pp. 1–16, 2017.

[189] S. Peng et al. Onboard observation task planning for an autonomous Earth observation satellite using long short-term memory. *IEEE Access* 6 (2018) 65118–65129.

About the authors

Yongming He

Received the B. S. degree in Logistics Engineering from Chang'an University, China, in 2014, and the M. S. and Ph. D. degrees in Management Science and Engineering at National University of Defense Technology (NUDT), China, in 2016 and 2021, respectively. He is also a visiting Ph. D. student at the University of Alberta, Edmonton, AB, Canada, from November 2018 to November 2019. Now, He is an associate professor at the College of Systems Engineering of NUDT, a member of 10th Young Elite Scientists Sponsorship Program by China Association for Science and Technology (CAST). Dr. He has 12 years of research and work experience in the field of intelligent task planning, optimization, and scheduling. He received two second prizes in Natural Science from the Hunan Province, one first prize in Natural Science from the China Simulation Federation, and one second prize in Natural Science from the Chinese Association of Automation, published over 20 papers, and authorized 12 Chinese patents.

Yingwu Chen

Received the B. S. degree in Automation, the M. S. degree in System Engineering, and the Ph. D. degree in Engineering from the National University of Defense Technology (NUDT), Changsha, China, in 1984, 1987, and 1994, respectively. Now, Professor Chen has been a distinguished professor and doctoral supervisor of the College of Systems Engineering, NUDT, a member of the 14th National Committee of the Chinese People's Political Consultative Conference (CPPCC). Professor Chen has been engaged in research in satellite intelligent task planning and complex system decision analysis for over 20 years, authored more than 70 academic articles, published 7 textbooks and monographs, and won two first prizes and five second prizes for the Provincial Science and Technology Award.

https://doi.org/10.1515/9783111585109-011

Index

https://doi.org/10.1515/9783111585109-012

www.ingramcontent.com/pod-product-compliance
Lightning Source LLC
Chambersburg PA
CBHW061412210326
41598CB00035B/6190